SUNWINGS

THE HARROWSMITH GUIDE TO SOLAR ADDITION ARCHITECTURE

BY MERILYN MOHR – ILLUSTRATIONS BY JOHN BIANCHI

Canadian Cataloguing in Publication Data

Mohr, Merilyn
  Sunwings

Bibliography: p.142
Includes index.
ISBN 0-920656-37-4

1. Solar greenhouses - Design and construction.
2. Solar collectors - Design and construction.
3. Solar houses - Design and construction.
4. Solar energy - Passive systems.   I. Title.

TH7414.M64 1985   697'.78   C85-099488-8

Trade distribution by Firefly Books, Toronto

Printed in Canada for:
Camden House Publishing Ltd.
7 Queen Victoria Road
Camden East, Ontario
K0K 1J0

Front Cover: Jeff Carroll
Back Cover: Jim Merrithew

This book is as precise and accurate as
experience, care and responsible
research could make it. However,
because of the variability of local
conditions, materials and personal skills,
neither the author nor the publisher
assumes any responsibility for any
accident, injury or loss that may result
from errors or from following the
procedures outlined in these pages.

This book is dedicated to my father, who
gave me a place to grow in the sun.

Acknowledgements

This book is based on the experience of dozens of designers, builders and homeowners across the country, who generously opened their design portfolios and their homes to public scrutiny. The author gratefully acknowledges their enthusiasm and invaluable assistance. She is especially indebted to Don Roscoe and the thorough greenhouse course developed by Solar Nova Scotia and the Nova Scotia Department of Mines and Energy; Michael Dorgan and Alan Seymour, who compiled *Construction Details for Attached Sunspaces* for the National Research Council; and Michael Glover, Michael Kerfoot, John Hix and Ross Rogers for their valuable comments. Michael Webster, Wayne Grady, Charlotte DuChene, Roberta Voteary and Linda Menyes were very helpful in seeing the manuscript through the press, and I would like to extend my personal appreciation to the indefatigable Miss Potts, of the North Bay Public Library, and to Dwayne Myers, guide to the electronic age.

Contents

Preface

A place in the sun

'Give me the splendid silent sun with all his beams full-dazzling'

— Walt Whitman

On a frigid February day, Rory and Christina Dowler of Russell, Ontario, lounge in their lawn chairs, basking in the hot afternoon sun and nibbling a salad of fresh-picked greens and tomatoes. Like hundreds of other snow-bound Canadians from Whycocomagh to West Vancouver, the Dowlers have found a way to bring Florida north for less than the price of a plane ticket. While their neighbours fly south at the first hint of cold weather, they simply step into their sunwing to enjoy the warmth and greenery of summer.

A "sunwing" is an addition that has been carefully designed and constructed to capture and hold the sun's radiation. It can extend from any kind of house — Victorian mansion or clapboard cottage — and can be as large or small as fancy and finances allow. Though it costs no more to build than a conventional addition, a well-designed sunwing can be virtually energy self-sufficient, exacting no payment for opening an otherwise ordinary house to the unique pleasures of the sun. Since he added a sunwing to the front of his Toronto house, Garnet McDiarmid grows orchids and camellias year-round; George Matthews of Dartmouth, Nova Scotia, starts thousands of seedlings in the spring, supplying himself and his neighbours with garden plants; Greg and Linda Powell can take a soothing dip in their hot tub in December, surrounded by lush houseplants and a spectacular view of the snow-covered foothills of southern Alberta; even as far north as Whitehorse, Wayne Wilkinson's sunwing supplies the family with home-grown greens most of the year.

For many people, sunwings are the best thing to happen to winter since the snow shovel. They bring a shaft of sunlight into Canadian homes when it is needed most. This country's climate has jokingly been characterized as "nine months winter and three months late fall," an exaggeration that comes uncomfortably close to the truth. The land wakes up slowly in the spring, enjoys a short frantic burst of growth in summer, then has a prolonged fade-out in the fall before the full dormancy of winter sets in. The winter itself is not long: it is the extended "shoulder seasons" that bring on acute cases of cabin fever.

"Winters here are long and cold," says David Quiring, a woodworker who added a sunwing to his central Saskatchewan log house. "It is often May or June before it is warm enough to sit outside and be comfortable. In our sunwing, though, it feels like summer when there's still snow on the ground."

A sunwing effectively shortens Canada's winter by incorporating principles of passive solar design and energy-efficient construction. The addition is oriented toward the south, its glazing positioned to collect as much of the winter sun's heat and light energy as possible. The building materials are chosen to absorb that warmth and keep it inside and are put together using construction techniques that minimize heat loss and maximize free solar gain. None of this adds significantly to total cost: it is just as easy to build on the south side of a house as on the north, and the windows are the same price whether they are concentrated in the south wall or distributed on three sides. The concrete that absorbs the heat is already part of the structure as is the air/vapour barrier that seals the warmth inside the room. What makes the difference is how these elements are arranged in the design and how they are detailed during construction, and that is what this book is about.

In most parts of the country, a

sunwing can be self-sufficient or use only minimal backup heat. As a bonus, it may slightly lower home heating costs by reducing heat loss through the wall it "buffers." Most homeowners report that although the sunwing needs an occasional heat infusion to keep it above freezing in the depths of winter, by February, it can collect enough heat from the sun to help warm the main house.

Canadian concern with home heating costs is a fairly recent phenomenon. The watershed for housing, as for transportation, was the formation of OPEC. Drivers quickly demanded cars capable of logging 40 instead of 14 miles to the gallon, but because houses are a larger investment, contractors and homeowners have been slower to incorporate similar changes in home design. In the last few years, however, energy upgrades have taken over an increasing share of the renovation market, as Canadians spend millions to stuff their attics and walls with insulation. In their single-minded attempts to improve thermal efficiency, many have ignored other approaches to home comfort. Remember the tale about the north wind and the sun vying to make the young man take off his coat? As the north wind blew, the youth only gathered the folds of his cloak more tightly around him, but when the sun gently shone, the youth willingly doffed his wrap. Likewise, though bundling a house in insulation offers some protection against the blasts of winter, a great measure of comfort can be gained by opening the house to the sun.

"The best thing," says Irene Shumada, proud owner of a northeastern Ontario sunwing, "is to go out there in T-shirt and shorts when the thermometer reads minus 22 degrees F outside and 80 degrees inside and just bask in the sun.

Not only does the heat feel good, but the sunlight itself is psychologically soothing." For many homeowners, such benefits outweigh the potential energy savings of a sunwing. Canadians have traditionally enclosed themselves in rooms with little more than a television-sized view of the world; the broad expanse of glazing in a sunwing offers a

wider vista and a more direct connection with the fundamental rhythms of nature. The sunwing itself becomes a massive sundial that marks the passing of the hours and the seasons as the sun's rays slant across the floor.

Tuning a house to its environment is not an avant-garde idea. In fact, it is only recently in human history that people have ignored such natural forces as sun and wind and have situated their houses

to accommodate sewer lines and civil servants. One of the first references to passive solar principles is a lecture by Socrates, in which he advises students to "build the south side loftier to get the winter sun." Religious structures from Stonehenge to Saint Peter's Church in Rome have been designed so that a single shaft of light at a certain time of year will illuminate a predetermined point in the room. On a more mundane level, this country's early farmers built their barns banked on the north and open to the south, with hay stored in the "attic" acting as insulation for the animal pens below. The farmhouses that squatted nearby often included a sun room — an unheated area that was closed off in winter but opened in early spring — as an intermediary indoor-outdoor room. The wide verandah typical of early Canadian homes was originally designed to shade the interior from the intense summer sun so unfamiliar to immigrants from the cloudy maritime climates of northern Europe. These verandahs also provided a place where ladies could sit and enjoy the outdoors without being exposed to the elements.

Sunwing design is rooted in such sensitivity to specific needs. There is no single "correct" passive solar addition that can be slapped on the side of any house. There are, however, basic principles of solar gain and heat-loss control that, once understood, can be adapted to suit a particular climate, property and house. The sunwing that results will not only be efficient, it will look like an integral part of the original structure, blending unobtrusively into a heritage home and improving the character of more modern buildings. Even a house that has a dozen clones in the neighbourhood can become unique with the addition of a sunwing.

A sunwing can be a greenhouse for growing winter food and flowers, a collector for supplementing the heating needs of the house, or a solarium for soaking up the midwinter sun, but most homeowners want some combination of these three functions. *Sunwings* focuses on passive solar additions that are designed primarily for winter-weary Canadians but that comfortably accommodate plants and effectively maximize solar heat. By explaining how function affects design, the book will help homeowners set personal priorities, guiding them step-by-step through the preliminary planning process – evaluating the solar potential of the climate, the site and the house itself. By looking at the layout and design of their houses, as well as the practical limitations of their properties, readers will be able to determine the best place to extend their houses to the sun. The design section that follows explains the elements of a sunwing and how they fit together to create an energy-efficient, sun-filled room. Armed with this knowledge, homeowners can realistically assess the merits of prefabricated sunwings (Chapter 5) or go on to design and build their own (Chapter 6). Rather than describe the step-by-step construction of an "average" sunwing, the last two chapters discuss the materials and construction techniques that apply specifically to passive solar additions.

The book contains no complicated charts or physics formulae that only a solar engineer can decipher. Instead, it is written from a practical point of view, giving rules of thumb to get the reader started in the design and construction processes. When the trendy architectural jargon is stripped away, sunwing design is mostly common sense. In fact, when Martin Foss of

Ashton, Ontario, found solar books too complex for his needs, he simply developed his own design that was as close to perfect as any that surfaced in our cross-country survey.

Sunwings is based on the results of that survey. Scores of architects, designers, builders and homeowners from Newfoundland to the Northwest

Territories were interviewed to discover firsthand what works for Canadian homes. As they described their sunwings, patterns began to emerge: certain accepted components of sunwing design, such as remote thermal storage and movable insulation, were not as cost-effective as expected; principles that made sense in theory, like natural convection loops to move warm air, did not necessarily follow textbook rules in

practice. Out of this pool of experience, *Sunwings* synthesizes the best, most current advice on the design and construction of passive solar additions. What makes it come alive are the real-life stories of the dozens of homeowners who enjoy their sunwings every day.

Some of the very best sunwings from the survey are profiled in "A Solar Sampler," a source of inspiration and information for anyone considering a passive solar addition. They prove conclusively that sunwings can be beautiful, practical and affordable. The case studies cover every climatic region of the country, most house styles and materials and a diversity of uses. The price tags vary from hundreds to many thousands of dollars, depending on the complexity of the design, the materials used and who did the construction. None are completely without problems, glitches from which readers can learn as they develop their own sunwing designs.

Not all readers will be do-it-yourself designer/builders, but even those who hire the job out can benefit from the information in these pages. Home-owners who are sure of what they want and know what is involved in the project will make better clients when they begin to talk to professionals. They will be more able to assess the advice of their designer or contractor and will be more realistic in their expectations of the people they hire and of the sunwing itself. For those considering buying prefabricated greenhouse or solarium kits, *Sunwings* will make them more discriminating buyers. The book focuses specifically on adding a sunwing to an existing structure, but it can be just as useful to those interested in incorporating a passive solar wing into plans for a new house. As a source of light, heat and greens, a sunwing is an incomparable addition to any house.

A well-placed sun room lends an airy feeling to a house, opening the home to the rhythms of nature. Norm and Doreen Hutton, Apsley, Ontario

1 Hatching a Plan

Form and function

'The proper choice of goals is half the battle.'

— William Shurcliff

Rory and Christina Dowler had several goals in mind when they designed their sunwing in the winter of 1981. They wanted a pleasant living space for sunny winter days and room to grow a wide variety of indoor plants, bedding plants in spring and the makings for fresh salads in winter. Furthermore, they wanted an energy-efficient addition that would not cost a fortune to heat. "As I built the addition, I had an image floating around in my head," recalls Rory. "I saw myself sitting in the sun on a Christmas morning surrounded by vegetation, eating a toasted tomato sandwich — the tomato and lettuce freshly picked."

That vision has materialized. All winter, the Dowlers eat fresh lettuce, radishes and tomatoes, and in spring, they start seedlings for the garden and flower beds. While the snow blows outside, they sit among trailing vines and blooming geraniums. In summer, they enjoy the cool night breeze out of range of mosquitoes. For many Canadians, however, their passive solar additions have been living nightmares: plants seared by day, heat sucked out of the house by night, inviting spaces for neither man nor magnolia. The Dowlers knew what they wanted in a sunwing and understood the design compromises necessary to achieve it; too many sunwing builders don't.

A sunwing can be many things — a heat collector, a greenhouse or a solarium where people bask in the sun. As for any building, the design of a sunwing is determined by its proposed use. Every structure is a climate-control unit, with roof, walls and floor combining to create a unique interior space that can be insulated from the conditions around it and then heated or cooled to become anything from a deep-freeze to a sauna. The artificial climate inside the enclosed space is created partly by the design of the structure and partly by the heating and cooling systems that are added to it. Exactly what form that inside climate takes depends on how the space will be used: while it may be obvious that cacti and watercress need vastly different environments, the same holds true for people and plants.

Each of the three main sunwing functions requires different and sometimes conflicting design features. A single sunwing *can* collect heat, grow plants and house the hot tub, but not without compromise: a true greenhouse will not be an efficient heat collector, and a solar furnace is certainly no place to serve afternoon tea. If heat is the goal, the sunwing's use as a greenhouse or living room will be restricted; if growing plants is the primary objective, sacrifices will have to be made in heating efficiency and livability; and if the space is designed as a sun room for people, growing conditions and heat collection will have to be adapted to human needs.

Therefore, the first and most crucial step in planning a sunwing takes place before a single piece of sod is moved — even before pencil meets drafting paper. The homeowner must ponder his or her priorities. If the sunwing does not suit the life style of the family it serves, it will not be used. A good professional designer spends a lot of time pinpointing the needs and expectations of the client, but this stage is too often neglected by a do-it-yourself designer, even though it is the one part of the process that the homeowner indisputably does best. No one can confront the issues of sunwing design as directly as those who have to live with its success or failure.

The illustrations on the following pages show how dramatically function affects the profile of the sunwing. A

Jim Merrithew

PLANNING FUNCTIONS When planning a sunwing, homeowners should always consider what its primary use will be.

Add significant heat to house
Serious food production
Grow specific houseplants
Be energy self-sufficient
Start seedlings
Grow general houseplants
Frost protection (wintering over)
Living space
Add marginal heat to house
Collector Sunwing
Greenhouse Sunwing
Solarium Sunwing
Integrated Solarium
Buffered Solarium

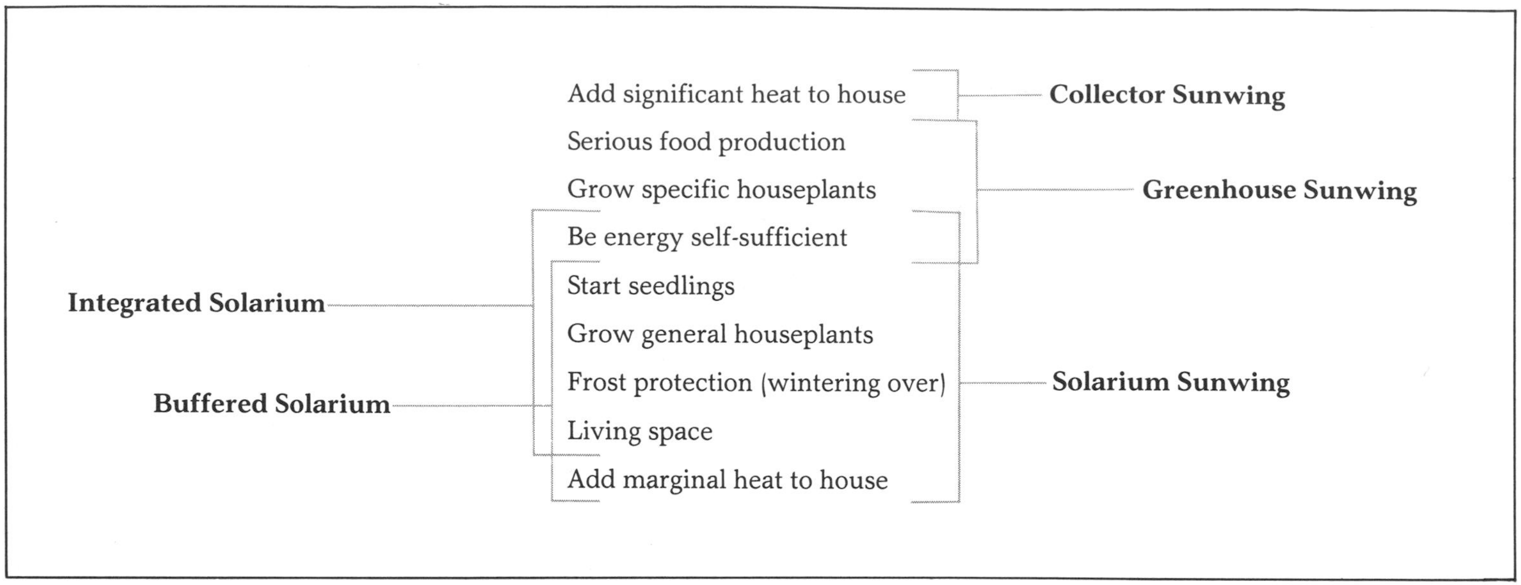

	Collector Sunwing	Greenhouse Sunwing	Integrated Solarium Sunwing	Buffered Solarium Sunwing
Backup heat	no	yes	yes	no
Movable Insulation	no	yes	yes	no
Shading	no	yes	yes	no
Thermal storage	no	yes	optional	optional
Glazing - slope	slope for max. heat	slope for light	vertical	vertical
south	max.	max.	max.	max.
east/west	no	yes	optional	optional
roof	no	yes	optional	optional
Interior finishes	heat-resistant dark	moisture-resistant light-reflective	low glare	low glare
Ventilation	to house	max.	optional	optional
Operator involvement	max.	max.	some	little
Controlled integration with house	yes	yes	no	yes

The sun room attached to this refurbished farmhouse (previous page) combines the vertical glazing typical of a sun room with the skylights needed for good plant growth. Martin and Betty Foss, Ashton, Ontario

sunwing devoted solely to **heat collection** is tall and narrow, with only enough floor space to create the correct angle for the collector. The glazing, oriented unerringly toward solar south, is perpendicular to the sun's rays when they are at their midwinter low so that the collector can gather maximum heat when it is needed most. Such a sunwing will likely be two storeys high to increase the collector area and to encourage warm air to rise so it can be immediately distributed to the house. Because a collector-sunwing is designed to heat the main house and not itself, it includes no heat-loss controls, thermal storage or backup heating system. Consequently, the temperatures in the sunwing itself swing wildly from the outside winter low to a sunny-day high of up to 150 degrees F. There may be a few hours in between when the space is habitable, but the conditions are too inconsistent for plants and too unpredictable for most people.

A sunwing intended as a **greenhouse** is designed to provide adequate light and heat for the kind of plants the homeowner wants to grow. Its south glazing, chosen for solar transmission rather than visual clarity, is sloped or continued overhead to provide maximum, even light. The addition is narrow, generally 8 to 10 feet deep, so that the winter sun can penetrate to the back wall, and it includes east and west wall glazing, allowing plants to be bathed in light from early morning to late afternoon. Although a high proportion of glazing is necessary for active growth, in a cold climate like Canada's, too much glass will create temperature swings that most plants cannot tolerate. To moderate the extremes, a greenhouse includes backup heat and movable insulation to counter nighttime heat loss, storage mass to

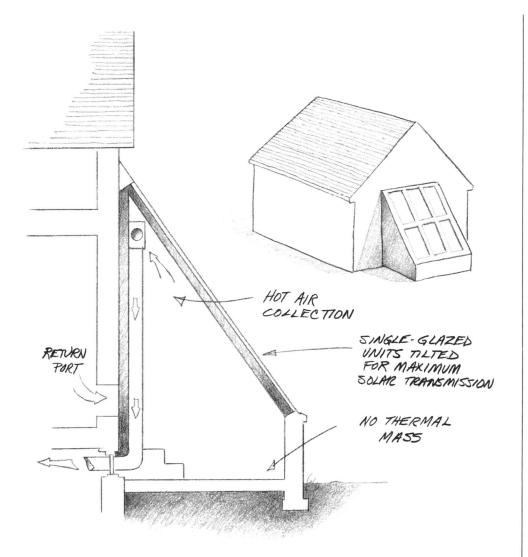

HOT AIR COLLECTION

SINGLE-GLAZED UNITS TILTED FOR MAXIMUM SOLAR TRANSMISSION

NO THERMAL MASS

RETURN PORT

absorb daytime solar gain, shade to protect the plants from intense summer sun, and ventilation to flush out excess heat and provide a fresh supply of carbon dioxide. The humidity inside a greenhouse is higher than is comfortable for humans and creates more dampness than conventional building materials can tolerate. For this utilitarian design, finishes are chosen not for aesthetics but

for their ability to reflect light, easy maintenance and resistance to rot.

A **solarium** is much less stringent in its design demands than either a greenhouse or heat collector. Instead of trying to maximize light or heat, a solarium-sunwing concentrates on collecting and retaining comfortable levels of solar heat. Instead of a sloped south wall, a solarium has vertical

GREENHOUSE SUNWING
Because climate control is
essential to healthy growth,
a sunwing devoted to
horticulture is the least
flexible in its design.

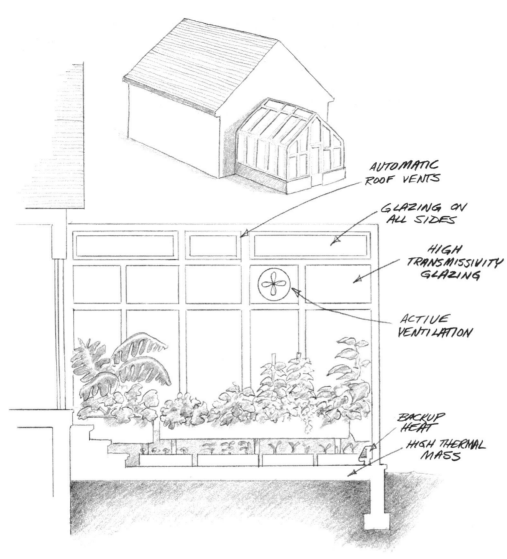

AUTOMATIC
ROOF VENTS

GLAZING ON
ALL SIDES

HIGH
TRANSMISSIVITY
GLAZING

ACTIVE
VENTILATION

BACKUP
HEAT

HIGH THERMAL
MASS

glazing that collects as much solar heat energy as a slope but loses less heat and is easier and cheaper to install. East-west and overhead glazings are eliminated in the interest of energy efficiency, and orientation can be modified to accommodate human rather than plant preferences.

As a primary function, there is no doubt that the solarium offers the most design freedom. Indeed, cold-climate designers generally agree that a sunwing designed exclusively for heat collection does not make economic sense for northern latitudes: a room-sized "solar furnace" is a large investment with only limited potential return, since a sunwing cannot hope to fulfill a home's heating needs. Likewise, a true greenhouse demands an investment of time and money that few homeowners anticipate. In Canada, the intensity and duration of light is so low around the winter solstice that grow lights are necessary to stimulate active plant growth. Because plants are dependent on their environment, a greenhouse must be automatically regulated with thermostatically controlled fans, vents and heaters to maintain growing conditions. According to Vancouver solar architect Richard Kadulski, the climate in southwestern British Columbia is sufficiently mild for plants to survive the winter without auxiliary heat, but the rest of Canada is not so privileged. Though the cost of operating a greenhouse depends on local climate, the crops grown and the type of backup used, it can be expensive: when an Ottawa family grew tomatoes in their sunspace, it cost as much to keep the addition at growing temperature as it did to heat the rest of their low-energy house.

However, heat collection is a viable *secondary* function for a solarium-sunwing. The vertical glazing promotes good solar gain, and the two-storey profile typical of a heat collector meshes well with human needs: the lower floor level remains a comfortable temperature, while hot air pools near the ceiling where it can be manually or mechanically vented into the main house. In midwinter, the sunwing will likely need backup heat to keep temperatures above freezing, but on an annual basis, depending on the design and use of the space, solar gain and energy consumption will usually even out, making the sunwing virtually energy self-sufficient.

"I always tell my clients not to expect any savings on their fuel bill when they build a sunwing," says Don Roscoe, Nova Scotia designer/builder. "But sometimes, there is a slight drop in overall energy consumption because the addition converts an exterior house wall to an interior wall." In other words, if a sunwing has a positive impact on home heating costs, it will probably stem more from its buffering effect than from solar gain. Even if the sunwing is unheated, its temperatures will not be as extreme as outdoors, so there will be less heat loss through the wall blanketed by new construction. If the common wall between the house and the addition is large and uninsulated, with ill-fitting single-glazed windows, the sunwing's effect on fuel bills can be significant.

A solarium really comes into its own, however, when plants are included in the plan. This, in fact, is what many people have in mind when they decide to build a "solar greenhouse" — a pleasant place to sit and somewhere to start seedlings and grow houseplants. According to Brian Marshall of Sun Shelters, a Toronto firm specializing in energy-efficient sunwing design and construction, "many clients *think* they want a greenhouse, but we've found that most people just want a table in the corner to start seedlings, maybe a hanging fern or two and a place to sit and enjoy the warm, bright winter sun."

Instead of designing a climate-controlled space for particular plants, species are selected that are flexible in their light and heat needs. There are many varieties that thrive on the relatively Spartan conditions of an unheated sunwing: jasmine needs a drop to as low as 35 degrees F to bloom profusely the next season; hydrangea, oleander and geraniums likewise enjoy the semi-dormancy induced by low temperatures; and winter-blooming plants like azaleas and cyclamens can adapt to temperatures close to freezing. John Hix even grows oranges in his

Primarily a solarium-sunwing, this addition was designed to combine first floor access with a second floor sundeck. Ron and Susan Alward, Mansonville, Quebec

Jim Merrithew

A buffered sunwing (left) is outside the thermal envelope of the house, and because it can be closed off, its temperature swings do not adversely affect the rest of the house. Sliding glass doors can be used to keep the sunwing thermally isolated but visually connected to adjacent rooms. Greg and Linda Powell, Priddis, Alberta

An integrated sunwing (top right) has no common wall to separate it from the main part of the house. Although this open concept encourages greater use of the space, there is less thermal control. Adjacent rooms may overheat in summer, and there will be more heat loss through the walls of this highly glazed room than through conventional walls. Ron and Susan Alward, Mansonville, Quebec

Although a solarium-sunwing (bottom right) is designed primarily for people, it can provide a healthy environment for plants as long as species are selected to match the conditions in the sunwing. Low-light crops like lettuce, as well as herbs and many flowering plants, thrive in the relatively cool temperatures of a buffered space. Rory and Christina Dowler, Russell, Ontario

Jim Merrithew

Roger Vernon

Jim Merrithew

sunwing north of Toronto, using baseboard heaters to keep ice off the watering cans but not to maintain temperatures at the level of human comfort.

"We've got to learn to give more appreciation to those plants that are willing to stick with us through the rigours of a Canadian winter – cool, low-light plants such as salad greens and herbs," says Alberta passive solar designer Michael Kerfoot. Indeed, many vegetables can survive the relatively cool, low-light conditions of a sunwing. In his unheated sunwing in Ayer's Cliff, Quebec, Luc Muyldermans grows witloof (French endive) under black polyethylene, which provides the darkness and warmth it desires. Aside from winter harvests of salad greens, celery, leeks and herbs, Muyldermans has found that cold-resistant vegetables brought in from the garden in late autumn can provide the family with fresh vegetables during the coldest part of the winter when light levels and temperatures are too low for young seedlings.

As the profiles of successful sunwings in "A Solar Sampler" prove, a well-thought-out master plan can produce an effective compromise – one with vertical glazing and good orientation for maximum heat collection, minimal overhead glazing for good light penetration, narrow built-in beds for plants, with lots of room left over for the Ping-Pong table. It will not grow plants as well as a nursery greenhouse or produce the maximum heat of a passive solar collector, but it will be light enough for healthy plants, collect enough heat to be self-sufficient and create an incomparable living space.

Unlike the single-purpose greenhouses and heat collectors, which are functional appendages to the house, there is a strong temptation to fully integrate this

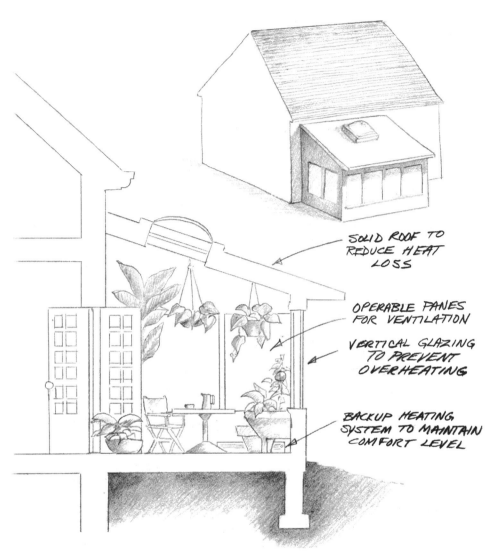

SOLID ROOF TO REDUCE HEAT LOSS

OPERABLE PANES FOR VENTILATION

VERTICAL GLAZING TO PREVENT OVERHEATING

BACKUP HEATING SYSTEM TO MAINTAIN COMFORT LEVEL

SOLARIU... Instead of ... heat gain or ... solarium desig... concentrates on ... comfort, combinin... principles of energy efficiency with passiv... solar collection.

"hybrid" sunwing into the living area of the house. With no barriers between old and new construction, heat could flow freely from sunwing to house when the sun is shining and from house to sunwing at night and on cloudy days. In a cold country like Canada, however, it is expensive to maintain consistent human comfort levels in a highly glazed structure. Even constructed to energy-efficient standards, an **integrated sunwing** – an addition that is not connected to the main part of the house by an insulated wall – will likely increase total heating bills as it drains house heat out through its large windows.

Originally, Martin Foss's sunwing was fully integrated with his renovated farmhouse outside Ottawa. "The first winter offered an opportunity to test a few assumptions. The common-wall opening was left uncovered, but the chill from the sunwing during cold nights was so noticeable that we decided to install pocket doors and insulate the remainder of the wall."

Foss thus created a **buffered sunwing**, an addition outside the thermal envelope of the house. The wall between the addition and the house is fully insulated, and the sunwing becomes a transitional indoor-outdoor space, more than a porch, less than a living room. The homeowner can use the solar heat when it is available but close off the space so that the house is not adversely affected when the temperature drops. The sunwing itself can swing with the weather or can contain heating and ventilation systems to counteract the heat loss/gain cycle. Aligning sunwing windows with house windows allows people to see out of and sun to shine into the main part of the house. At the same time, the sunwing creates a thermal break between the house and outdoors. Because the temperature differential across the buffered wall is thus narrowed, there is less heat loss from the house.

"This buffered sunspace offers a new kind of opportunity and life style that is hard to imagine," says John Hix, an award-winning cold-climate architect who considers the buffered sunspace to be archetypal for Canadian conditions. "The space is flexible, never a bad space, always usable — as a games room and mudroom after cross-country skiing, as a greenhouse for starting seedlings, as a summer sleeping porch."

Being outside the thermal envelope need not limit a sunwing's use. Doreen and Norm Hutton spent $2,000 to add a 200-square-foot sunwing to their central Ontario retirement home, an investment

that has brought them daily pleasure. "For us, it is a breakfast room, a station for bird watching, a sanctuary for our plants (23 of which are hanging) and, at the appropriate season, a place to force bulbs, start tubers and seeds, while at all times nurturing those plants lifted from last summer's garden. We count our blessings!"

That kind of satisfaction stems from sound planning. The homeowner should start by taking an honest look at the house to determine if its structure and condition merit the expense of adding a sunwing. Is a passive solar addition really the top priority, or would the family's resources be better spent on overdue maintenance or upgrading insulation levels? A sunwing will not do much to improve the comfort of a house that is underinsulated or leaking air like a sieve.

Once the decision is made to go ahead with a passive solar addition, priorities must be set. Make a list of everything the sunwing will be used for, then number the functions from most to least

important. Though that sounds very easy, it will likely take some soul-searching to come to grips with the fundamental purpose of the space. Once the primary function has been established, each of the secondary functions can probably be incorporated into the design with varying degrees of success. Beside each one, jot down the design features required for the activity. For instance, if the sunwing will contain planting beds, the homeowner may want an outside entrance or a special potting area. Finally, establish a rough dollar figure the family is willing to invest in a sunwing.

This initial stage can take days, weeks, months or even years, with the sunwing evolving as family interests change. In the meantime, keep a file of magazine clippings of promising design features, and visit home shows to find out about new products. "Have a friend build one first and help out," adds Eric Darwin, who learned much from helping a friend through the design and construction of her sunwing.

A good way to understand the implications of the final decisions, aside from reading the rest of this book, is to visit as many sunwings as possible. Talk to people who live with passive solar additions to discover what works and what doesn't in a particular locale. Next to politics and religion, the family castle is a favourite topic of conversation, and probably as controversial.

"Go to open houses for a year of Sundays," advises Darwin, who developed the design and did most of the construction on his own tri-level sunwing. "Go on solar and neighbourhood tours, visit new home developments that feature sun rooms, and any time you notice one on a house, leave a note asking to see it. Most people love to give tours — we do too!"

Jim Merrithew

SKYLIGHTS

EXHAUST PORT

HOT AIR COLLECTION DUCT

OPERABLE PANES FOR VENTILATION

SEED BEDS

BACKUP HEATING SYSTEM

RETURN PORT

INSULATED SLAB COVERED WITH CLAY TILE

COOL AIR RETURN

19

2 Spreading a Wing to the Sun

Gauging solar potential

'The sun, it shines everywhere.'

— William Shakespeare

When Wayne Wilkinson decided to add a sunwing to his house in Whitehorse, he found that Grey Mountain to the east blocked the winter sun until late morning, a neighbouring house shaded the site in the afternoon, and instead of facing south, his house tilted sharply to the southeast. At that latitude, the winter sun barely rises above the horizon, and when it does, its rays have to filter through a haze of wood smoke that persistently hangs over the valley. Nevertheless, Wilkinson overcame these seemingly insurmountable obstacles and built a sunwing that feeds his family year-round, saves $425 on his annual heating bill and is "a warm, sunny retreat from the grasp of winter."

As Wilkinson proved, a sunwing can be added successfully to any house as long as its design takes into account the idiosyncrasies of the local climate, the site and the house. The amount of solar heat and light energy the sunwing collects depends entirely on how many hours of sunshine the region receives, the sunblocks that shade the property and the orientation of the sunwing itself.

ORIENTATION

Because Canada lies in the northern hemisphere, windows have to face south to catch the sun. Yet the sun is constantly on the move, or so it seems. As the Earth spins, the sun tracks across the sky from east to west on a daily cycle, and from low on the horizon to high overhead and back down again on a yearly cycle. It rises and sets at its most northerly points on the summer solstice (on or around June 21), and on that day, the sun is as high overhead as it gets. By the winter solstice (around December 21), the sun rises and sets at its most southerly points on the horizon and is at its lowest zenith. The spring and fall equinoxes, around March 21 and September 21, mark the midpoints of this annual cycle.

In order to catch all the available solar radiation, a window would have to pivot with the sun, a feasible solution for active collectors but not for sunwing glazing. The next best thing is to face the glazing toward the midpoint in the sun's arc across the sky. That midpoint occurs at solar noon, exactly halfway between sunrise and sunset. The sun's position in the sky at solar noon is called solar, or true, south. A window oriented precisely toward solar south collects the most light and heat energy available to a fixed surface.

How close sunwing windows can come to this ideal depends to a large extent on the relationship between the house and the sun. Unfortunately, because solar orientation was not a priority when most Canadian houses were built, few houses will be perfectly oriented for a sunwing. To figure out where the house lies in relation to solar south, establish which wall faces most nearly south (look for windows that admit direct winter sun). The exact orientation of that wall can easily be determined by using a compass (see the map on page 24).

If there are no windows or doors in the south wall, the homeowner will have to go outside to locate solar south. In an unshaded area of the yard, about 10 feet out from the south side of the house, drive a 6-foot stake into the ground, checking with a level that it is vertical. (In spring and summer, the shadow may be too short to give an accurate reading, in which case, a plumb bob suspended from a tree or clothesline may produce enough cast shadow.) At exactly solar noon, the cast shadow indicates the solar north-south axis. Use stakes to mark the

Jim Merrithew

Because this sunwing is used for late-afternoon dips in the hot tub, it includes west glazing. Overheating is counteracted by shades and a thermal storage system that dumps excess solar gain into 50 cubic yards of rock buried under the floor. Greg and Linda Powell, Priddis, Alberta

Although this two-storey brick house has excellent solar orientation, a warehouse wall to the west blocks much of the afternoon sun in winter. By raising the back half of the sunwing roof an extra 4 feet, the homeowner was able to significantly extend his solar "day." Eric Darwin and Frances Dubois, Ottawa, Ontario

Perched on a rugged, south-facing slope (previous page), this sunwing is fully glazed on the south and east walls, but a main floor sundeck shades the west wall. Norm and Doreen Hutton, Apsley, Ontario

Roger Vernon

Jim Merrithew

beginning and end of the shadow, then extend the line to the wall of the house. The angle between this north-south line and a line perpendicular to the wall indicates how much the house deviates from solar south.

Fortunately, orientation is not a solar-south-or-nothing proposition. The house, and thus the sunwing, can face within 30 degrees of solar south and still garner 90 percent of the potential solar energy. Most Canadian designers recommend orienting passive solar structures within this 30-degree limit, though some consider even as much as 45 degrees off south acceptable. At that angle, a quarter of the available solar energy is sacrificed, so it is not recommended for northern latitudes or where heat gain is a priority.

Although a certain measure of play is possible without abandoning solar principles, the amount of deviation from south will affect the time of year the sunwing gets the most sun. The east and west walls of a house receive two to three times more solar energy in summer than they do in winter. Thus, the farther east or west a sunwing faces, the more winter solar gain is reduced and summer solar gain is increased.

Whether a sunwing faces east or west of south also affects the time of day that it receives the most sunshine. A western orientation exposes the sunwing to less morning sun, which is not particularly good for plants: in the low-light winter months, they need direct sunlight as early in the day as possible to renew photosynthesis. Therefore, greenhouse-sunwings that face more than 15 degrees west of solar south should include east wall glazing to provide enough light for active growth. However, for people who are late risers or who want a bright, sunny space after a hard day's work, the afternoon sunshine of a west-facing sunwing may be welcome.

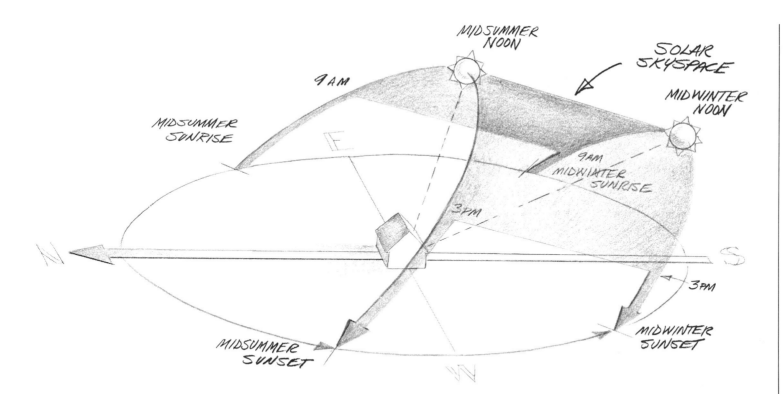

Labels on diagram: MIDSUMMER NOON, 9 AM, MIDSUMMER SUNRISE, SOLAR SKYSPACE, MIDWINTER NOON, 9 AM MIDWINTER SUNRISE, 3 PM, 3 PM, MIDSUMMER SUNSET, MIDWINTER SUNSET, N, E, S, W

It is important to remember that the days are as long in April as in August and that while evening warmth may be useful in spring, in midsummer, the sunwing will be getting maximum sunshine at the time of day when the outside air temperature is still high. In places like Whitehorse, where the heating season is 12 months long, this extra shot of late-afternoon sun may be appreciated, but in most of the country, it will cause serious overheating. If the sunwing faces west of solar south, summer shade and ventilation are particularly important. A windbreak should also be considered if the prevailing winds come from that direction (as they do in much of Canada).

Given the choice, most designers would rather tilt a sunwing east than west. An east-facing sunwing avoids the hot afternoon sun in summer and admits light early in the day when it is most appreciated, often by people as much as by plants. This is the ideal orientation for a sunwing that doubles as a breakfast nook and is especially appropriate for a sunwing designed as a season stretcher. In spring and fall, the sunwing warms up early, replenishing the heat lost during the cool night. By late afternoon, ambient temperatures are usually high enough to preclude the need for solar gain.

Personal preference for morning or afternoon sunshine will be limited by the orientation of the house. Although every house will offer several potential sunwing sites, one with an east-west axis – and consequently a long south wall – provides the most choice. If the longest dimension of the house runs north-south, there will be fewer options, but the biggest challenge is a house that sits at a 45-degree angle to south. This was the situation that faced George Dewar of St. Catharines, Ontario: only the corners of his L-shaped bungalow faced the sun at solar noon. His solution was a sunwing that wrapped around the corner, flattening the south-facing corner with two large trapezoidal windows.

SHADING

Facing windows in the right direction is no guarantee, however, that the sunwing will be solar-heated. Clouds in the sky and obstructions here on Earth can come between the sun and the glazing, blocking direct radiation.

Throughout the heating season, the sunwing should have a clear "solar

23

Solar south is not the same as magnetic south on the compass. In fact, as this map indicates, solar and magnetic south coincide only along a line of zero variation that runs roughly from Thunder Bay to Bathurst Island. To the east of that line, magnetic south is increasingly farther east of solar south, and likewise, to the west of zero, it is increasingly farther west. To find solar south, take a reading from the most south-facing wall of the house. Use a rectangular-based compass, making sure it is level, with no metal nearby to distort the reading. The number of degrees indicated will have to be adjusted for the variation between solar and magnetic south at any specific location. If you live east of the zero line, subtract the appropriate ''degrees of variation'' from the compass reading. Those to the west should add the degrees of variation. The final figure will indicate the orientation of the wall with respect to solar south (180 degrees), due east (90 degrees) and due west (270 degrees). For instance, the compass reading from the south wall of a house in Quebec City is 210 degrees. According to the map, the difference between solar south and magnetic south at this location is 19 degrees. Since Quebec lies east of the line of zero variation, the solar orientation of the house is 210 − 19 = 191 degrees. (Homeowners west of zero would add the degrees of variation.) Therefore, the wall faces 11 degrees west of solar south (191 − 180).

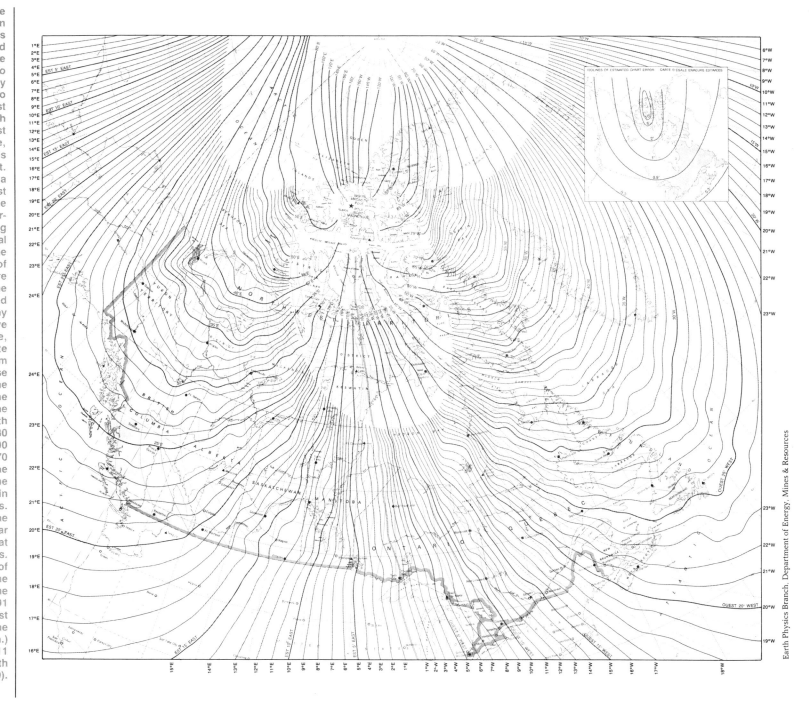

Earth Physics Branch, Department of Energy, Mines & Resources

window" during the midday hours, since 90 percent of the day's available solar energy strikes a site between 9 a.m. and 3 p.m. At the very least, the glazing should be in full sun during the winter months between 10 a.m. and 2 p.m., when the rays are most direct. In summer, shade will be welcome, but in winter, it can significantly reduce the sunwing's solar potential.

Trees are a common source of shading. Conifers block the sun year-round, but deciduous trees have always enjoyed a measure of solar acceptability because their life cycle seems in tune with passive solar design — providing shade in summer and obligingly shedding leaves in winter so the sun can peek through bare branches. The fact is, however, that even a leafless tree can have considerable shading impact. Research at Pennsylvania State University found that winter shading may be only 10 to 30 percent less than in summer, depending on the variety. Some species, like oak, hold onto their leaves well into the heart of winter, while others leaf out early in spring, shading the glazing before the heating season ends.

Trees directly in front of the proposed addition will cause the most winter shading, though lower branches can be pruned so that the top provides summer shade when the sun is high and only the trunk blocks the low winter sun. If a tree grows close to the sunwing but to the side, it may provide valuable summer shade without blocking winter sun at all. Bear in mind that trees close to the sunwing can also constitute a hazard by dropping branches, ice or fruit onto the glazing.

The Wilkinsons willingly cut down a backyard stand of small trees to give their sunwing a wider solar window, but when Gerald and Joan Donnelly built

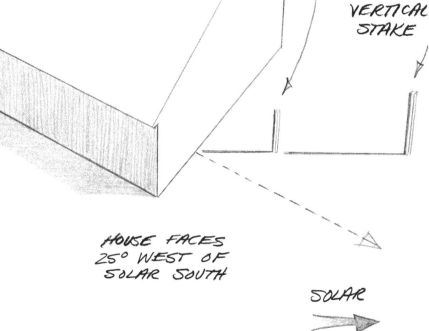

SMALL STAKE AT END OF SHADOW AT SOLAR NOON

6 FOOT VERTICAL STAKE

HOUSE FACES 25° WEST OF SOLAR SOUTH

SOLAR SOUTH

FINDING SOLAR SOUTH
At solar noon, the shadow cast by a vertical post indicates solar north and south, and if it strikes the house at a right angle, that wall is perfectly oriented for a sunwing. Most houses, like this one, face slightly east or west of solar south.

their addition, they did not even consider felling the three towering spruces in their front yard. Originally, the conifers were planted to mark the corners of the first homesteader's log cabin, but one now grows within 6 feet of the southeast corner of the glazing. Great care was taken during excavation to disturb its roots as little as possible, even though the tree blocks most of the low winter-morning sun. "No amount of extra heat we would get could compensate for the loss of these trees,"

says Joan. Aside from their historic and sentimental value, the trees provide cover for flocks of cardinals, jays, sparrows and grosbeaks that frequent the Donnelly feeding station just a few feet from the sunwing window.

When looking for potential sunblocks on the south side of a house, do not forget that trees and bushes grow. Low trees that seem insignificant at construction time could eventually shade the sunwing. If looking at the site in summer, keep in mind that shadows

SOLAR ORIENTATION
If a wall faces within 30
degrees of solar south, it
still enjoys 90 percent of the
available sunshine.
However, both the time of
day and the time of year
when the sunwing receives
the most sun are affected
when the sunwing is turned
to the east or west.

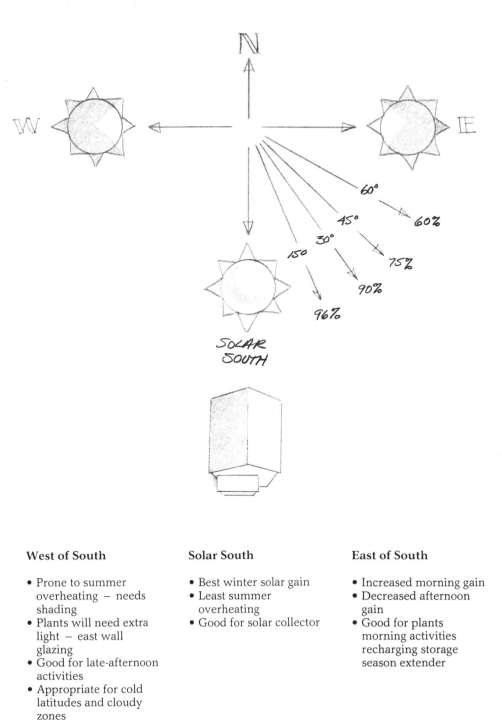

SOLAR
SOUTH

West of South

- Prone to summer overheating — needs shading
- Plants will need extra light — east wall glazing
- Good for late-afternoon activities
- Appropriate for cold latitudes and cloudy zones

Solar South

- Best winter solar gain
- Least summer overheating
- Good for solar collector

East of South

- Increased morning gain
- Decreased afternoon gain
- Good for plants morning activities recharging storage season extender

also grow longer as the days get shorter. For instance, a 15-foot-high shed in Kirkland Lake, Ontario, casts a shadow that is 8 feet long in summer but stretches to 35 feet in winter. The length of the shadow depends not only on the time of year but also on the latitude. On December 21, a 25-foot poplar would cast a shadow 60 feet long in Port Maitland, Nova Scotia (44 degrees north latitude). In Saskatoon (52 degrees north latitude), the same poplar would cast a 96-foot shadow, and in Peace River, Alberta (56 degrees north latitude), the shadow would extend 134 feet. In the Yukon, where the winter sun barely crests the horizon, if there is so much as a bush between the house and the horizon, solar radiation is blocked. As a general rule for most populated parts of Canada, any object south of the sunwing should be two to three times its height away from the south face, or it will shade the glazing in winter.

For a clear solar window in winter, there should be no obstructions within 45 degrees of the south glazing. The effects of shading can be determined simply by observing the proposed sunwing site over the course of a year. If the sunwing is to extend directly from a south wall, note the shadow patterns on that surface at various times of the day throughout the year, marking them on the wall or drawing a rough sketch. The potential sunwing site can also be photographed on a sunny day around the 21st of March, June, September and December.

Although informal observation of shading patterns is quite adequate, a professional site evaluation can be done by a solar specialist or by the homeowner using one of the solar manuals listed in "Sources." Technical confirmation is best provided by a sunpath chart, a two-dimensional representation of the three-dimensional movements of the sun in the southern sky. It shows the sun's approximate position for any hour on certain days of the year and the shadows that are produced when it is blocked by objects on the site. A sunpath chart takes time and patience to prepare and is important only if the sunwing's main function is solar collection. Another evaluation device is a solar shading mask. When held up to the southern horizon, it shows at a glance the sun's path through every season and how solar radiation will be blocked by man-made and natural obstructions.

Few sites will be completely in the clear, but there should be no more than 20 percent shading of morning or afternoon sun and no more than 10 percent between the critical hours of 10 a.m. and 2 p.m. If the ideal attachment site is heavily shaded, simply moving the proposed addition a few feet may bring it into the sunlight. A two-storey brick warehouse wall just 10 feet from the west side of Eric Darwin's house shaded his proposed sunwing site for most of the afternoon. He circumvented the problem by going up, extending his sunwing with a clerestory that gives him an extra hour and a half of sunlight on winter afternoons.

The sunwing can be designed with due consideration for existing houses and sheds in its sunpath, but the homeowner has little control over future construction that could put the extension in the dark. A Scarborough, Ontario, couple had the heart-breaking experience of watching a four-storey apartment building rise just 20 feet from their property line, completely shading their brand-new sunwing. This is one of the risks of building a passive solar structure in a country that has no laws to guarantee solar access. The best a homeowner can do is consult neighbours about their future building plans and hope that none sells out to a developer.

CLIMATE

A sunwing might be oriented due south with not a shed or shrub in sight and still not get enough solar radiation to grow plants or to keep the furnace from cutting in. Even the most carefully positioned sunwing must depend on the vagaries of local climate for the amount of sunshine it receives.

There is an Inuit legend in which a young hero spends his nights rowing Goddess Sun from west to east, saving her strength to rise again next morning; and, in fact, in the northern parts of the country, the winter sun shines so weakly and rises so half-heartedly that it seems a frail deity indeed. The northern territories see the greatest fluctuation in temperatures in a country noted for its extremes: winters are long, dark and cold, with the thermometer dipping below minus 30 degrees F for days on end, and frost and snow can be expected 12 months of the year; yet Fort Smith of the Northwest Territories has recorded highs of 102 degrees.

Only a small proportion of the population lives within the frigid zone, where the sun stays below the horizon for at least one full day a year; most Canadians make their homes within the temperate zone, where the sun is never directly overhead and yet never fails to show its face. Ironically, those living in the southernmost parts of the country do not necessarily get the most sun. Canada's own lotus land, British Columbia, enjoys the mildest climate but is capped with cloud cover 7 days out of every 10. The mean number of hours of bright sunshine for Vancouver in December is only 38, whereas Swift Current, Saskatchewan, records 87. Warmed by the Pacific Ocean and protected by the Rockies, British Columbia's climate resembles that of northern France. The coast is prone to fog, especially in the fall: 80 days of the year, Vancouver is blanketed in fog compounded by industrial haze.

East of the Rockies, the large land mass of the Prairies is subject to a

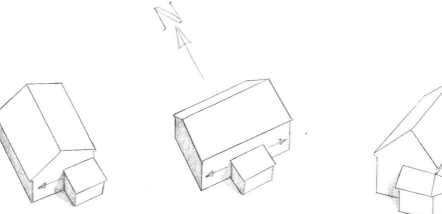

NORTH/SOUTH AXIS   EAST/WEST AXIS   ANGLED AXIS

SOLAR SITE OPTIONS
The number of potential sunwing sites depends partly on the axis or longest dimension of the house. An east-west axis, centre, offers the most design freedom; an angled axis, right, presents the greatest challenge, since only the corner of the house faces south.

SHADOW LENGTH
In Nova Scotia, this spruce tree would have to be almost three times its height away from the sunwing to prevent shading, but in northern Alberta, it would have to be more than five tree-heights from the south glazing.

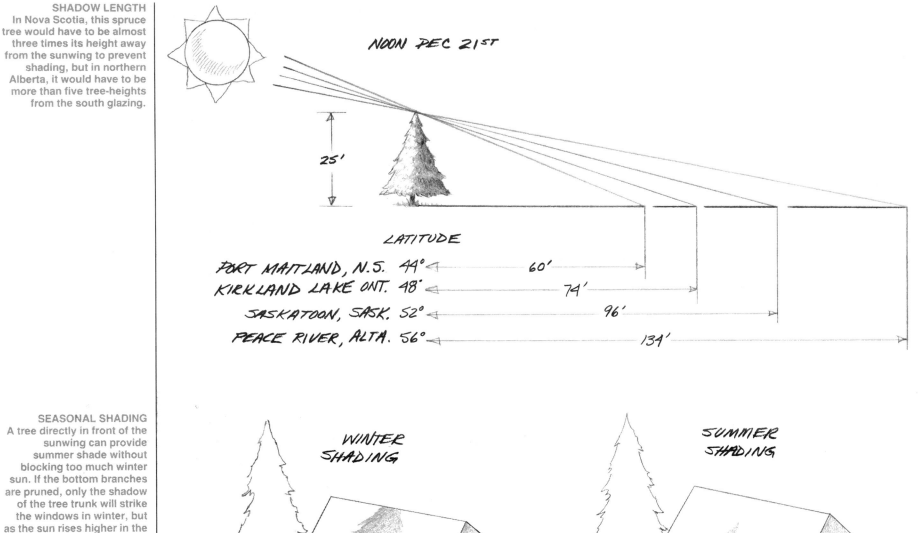

NOON DEC 21ST

25'

LATITUDE

PORT MAITLAND, N.S. 44° —— 60' ——
KIRKLAND LAKE ONT. 48° —— 74' ——
SASKATOON, SASK. 52° —— 96' ——
PEACE RIVER, ALTA. 56° —— 134' ——

SEASONAL SHADING
A tree directly in front of the sunwing can provide summer shade without blocking too much winter sun. If the bottom branches are pruned, only the shadow of the tree trunk will strike the windows in winter, but as the sun rises higher in the sky, the crown will shade the glazing.

WINTER SHADING

SUMMER SHADING

continental climate characterized by dramatic changes in temperature from winter to summer and from day to night. Because the weather in this area is determined by the dry westerly winds from over the Rockies and the cool dry Arctic air from the north, winters are very cold, but they are also sunny: Alberta gets 80 percent as much winter sun as Florida. Almost invariably, when outside temperatures plunge to minus 40 degrees F, the sky is clear, providing excellent solar gain when it is most needed. Even Quebec Premier Ministre René Lévesque, visiting Regina in February, was heard to enthuse about the "nice, clean Prairie cold."

The warmest part of the country, next to the Pacific Coast, is southwestern Ontario, which lies at Canada's lowest latitude. Moderated by the Great Lakes and the warm humid air swooping up from the Gulf of Mexico, this southern tip of the country enjoys a modified continental climate — less extreme temperatures than the Prairies and less precipitation than the coasts. The Maritimes, sitting on the lee side of the continent, also enjoy a relatively warm, mild climate, though subject to fog and cloud (Annapolis Royal has a mean of only 48 hours of sunshine in December). Newfoundland is both colder and cloudier than its southern neighbours, especially on the East Coast: there is fog in St. John's one out of every three days of the year.

The number of hours of bright winter sunshine typical of an area determines how much heat the sunwing generates, but even on cloudy days, some solar energy can filter through what seems to be a complete sunblock. The sun's radiation is broken into three types: direct, diffuse and reflected. Direct radiation is the most intense, but even this can be boosted by radiation reflected

Towering conifers that mark the site of a homesteader's cabin block some solar gain in winter but were left standing because of their historical and aesthetic value. Gerald and Joan Donnelly, Albion Hills, Ontario

Gail Harvey

from bright surfaces such as sand or snow. When direct radiation is scattered by dust and water molecules, it becomes diffused and less intense yet still provides some useful heat. On a clear day, the amount of direct solar radiation striking a surface can be as high as 365 BTUs per hour per square foot. Even with a dense cloud cover 4 to 5 miles thick, the diffuse radiation will still amount to about 21 BTUs per hour per square foot. (A bright overcast day will be somewhere between these two extremes.) This diffuse radiation can be used to advantage. For instance, a homeowner in Revelstoke, British Columbia, in the rainshadow of the Rockies, where there is a preponderance of diffuse radiation in summer, can design a sunwing with sloped glazing or

a westerly orientation without worrying so much about overheating.

The local weather office should be able to provide specific details as to record lows and highs, average monthly temperatures and the average number of hours of sun that can be expected during the heating season. However, to paraphrase an old adage, climate is what you expect, weather is what you get. Topography, landscaping, and proximity to surrounding buildings, hills, lakes, trees and open fields inevitably shape the regional weather into a unique microclimate. A house downwind from a factory may receive less direct solar radiation than is typical of the region because sunlight is diffused by air pollutants. Large bodies of water can produce early-morning fog or localized

cloud cover and storms. Don Roscoe built his own house on a point of land near Peggy's Cove, Nova Scotia. Although the coast is often enveloped in early-morning fog, his windows have clear solar access because the breezes sweeping across the point keep the fog away from his shores.

Meteorologists can predict fairly accurately how much solar radiation will be available over a wide area during an average December, but it is next to impossible to foretell how much sun there will be at a particular address next January, or even next Tuesday. Passive solar systems, however, are site-specific, sensitive to local conditions, and homeowners who ignore this aspect of planning risk great disappointment. In the summer of 1984, Joan Tucker added a prefab solar greenhouse to her coastal Newfoundland house. After spending almost $10,000 on the addition, she realized that her corner of the country is blanketed with a heavy cloud cover most of the winter: "I've lived here for over 20 years, and I guess we never did get much sun in winter. But you know, I never noticed it until we built the greenhouse and needed the sun!"

The homeowner cannot assume that once a sunwing is built, there will somehow be enough sun to make it a useful space. The site must be observed carefully so that the sunwing design can be adapted to local weather. It might be a good idea to start the process in the fall, taking weather notes during the heating season as the design evolves and matures. In many ways, homeowners considering additions have an advantage over new-home builders − since they already live on their sites, they can faithfully observe immediate conditions before preparing their final designs.

Though an energy-efficient addition can be built to suit any climate,

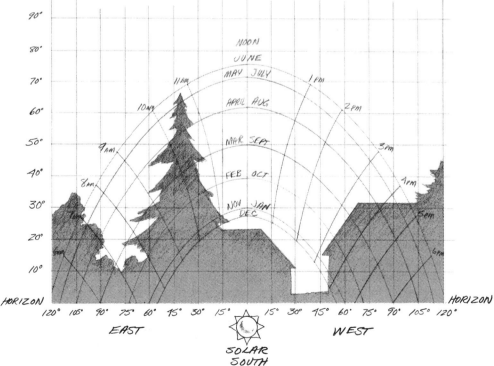

Robert Tinker

SOLAR POTENTIAL
1. Draw a rough sketch of the property. 2. Draw in an overhead view of the house, roughly to scale and oriented correctly. 3. Find solar south, and mark it on the sketch. 4. Draw in any buildings or trees that might shade the southern part of the property. Remember to consider tall structures that lie beyond the property line. Lightly fill in the areas covered by shade in the winter. (Summer shade is an advantage, not a liability, so it does not need to be indicated.) 5. Mark wind direction and any particular idiosyncrasies of the site's microclimate. 6. Areas along the south-facing wall of the house that are not shaded in winter represent potential sunwing sites.

homeowners who observe local conditions will have more realistic expectations. If a microclimate yields only three days of sunshine in December, the sunwing will not likely ripen tomatoes at Christmas without artificial grow lights. If fog consistently blocks morning sun, the sunwing can be oriented slightly southwest. The number of sunny days between October and March should provide a fairly good idea of how often an unheated sunwing can be used or how often backup heat will be needed.

How those sunny days are arranged is also important. Are they scattered equally throughout the month, or is the area subject to week-long cloud cover followed by week-long sunny spells? If the sunwing has regular infusions of sunshine, it will more likely stay up to temperature than if the space has to warm up after a prolonged cold spell that has drained all of its stored heat. The importance of exposure to solar heat energy is confirmed in the seasonal cycles of the Earth itself. On March 21 and September 21, the Earth receives the same amount of radiation, but because the atmosphere has been progressively warmed throughout the summer, September invariably seems closer to summer than March. How long the sunwing is exposed to the sun's warming rays will determine how summerlike the room will be, even in the middle of winter.

In the Yukon, winter hangs on like a bad cold. "Minus-30-degree temperatures have been known to persist into April," says Wilkinson. Fortunately, April and May are also the sunniest months in the Far North, with sparkling clear skies day after day. Though the ground outside remains frozen under a blanket of snow, inside the Wilkinson sunwing, spring arrives on schedule.

3 Connective Issues

House and site analysis

'Architecture is a backdrop for worthwhile human activity.'
— Raymond Moriyama

As a model of sun-oriented design, the solar greenhouse attached to Toronto's Ecology House is less than perfect. Because the sunwing extends from the west side of the energy demonstration house, its north wall is bared to the weather, the outside door opens into winter winds, and there are no east windows to give extra light to sun-hungry plants. Yet given the layout of the house and local setback allowances, it was the only attachment site that effectively combined good solar gain with convenient access to the kitchen. Though not ideal from an energy perspective, the position of the sunwing is nevertheless appropriate for the property and meets the needs of those who use it.

Practical and personal concerns such as these ultimately take precedence over solar principles. Although several sites on a house may offer a good window to the sun, many will be eliminated on legal, practical or aesthetic grounds. By examining the property and the house, the homeowner will be able to choose the very best place to extend a sunwing.

THE PROPERTY

In many localities, municipal officials will have a hand in where the sunwing is built. Local zoning regulations spell out in detail what a particular piece of property may be used for, how tall buildings may be, how far they must be set back from the street and from property boundaries, how much side clearance must be maintained between neighbours, and what proportion of the property can be covered by buildings. If solar sites are limited, the homeowner can apply for a "variance," which allows him to break the rules provided that no one's rights are infringed. According to Bob Argue, author of *Plans, Permits & Payments*, more than 80 percent of

Ontario requests are approved, which suggests that if the request is reasonable and unchallenged by neighbours, it probably will be granted.

These kinds of regulations are often more rigidly enforced in urban and suburban areas than in rural townships and villages. Because they differ substantially across the country, only one generalization holds true: It is the homeowner's responsibility to know the legal constraints on expanding his house and to abide by them scrupulously.

Although a sunwing seems an intensely personal and sensible piece of construction, in urban and suburban areas, there may be restrictive covenants that prevent certain changes to the house. In some municipalities, neighbours may have a say in the design. When Garnet McDiarmid decided to replace his decrepit front porch with a sunwing, he distributed copies of the plan as required. Although there seemed to be no objections, on the day of the hearing, his next-door neighbour informed him that he would oppose the project. McDiarmid withdrew the application, and when he reapplied two years later, the objection was overruled by the otherwise enthusiastic neighbourhood support.

Early consultation with neighbours can uncover other useful information, such as future building plans that might affect sunpaths. If the woman next door is a long-time resident, she may know if any easements prohibit extending the house in a certain direction. An easement gives another person the right to use part of the property. This can be a right-of-way for a utility company to maintain cables or water mains, or in rural areas, it may give access to a trapper or lumber company. There can be no construction on land that has

Gail Harvey

SITING A SUNWING
Locations with lots of
sunshine may have to be
sacrificed because of zoning
bylaws and practical
obstacles like septic
systems.

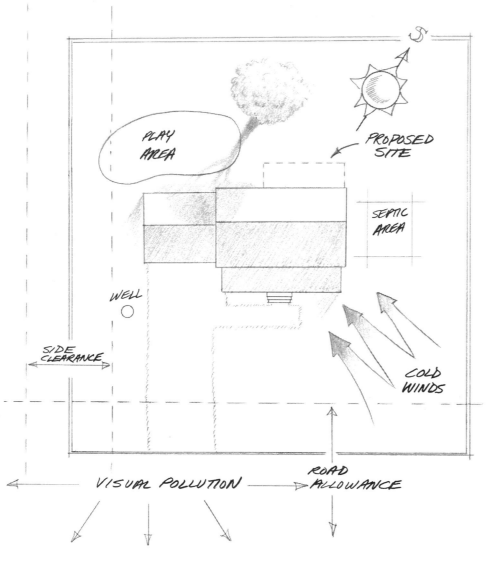

SIDE CLEARANCE

WELL

PLAY AREA

PROPOSED SITE

SEPTIC AREA

COLD WINDS

VISUAL POLLUTION

ROAD ALLOWANCE

Although rural municipalities
often place fewer regulatory
restrictions on sunwing
builders than on their urban
counterparts, rural sites are
still subject to the
constraints of the house and
landscape (previous page).
Gerald and Joan Donnelly,
Albion Hills, Ontario

become an easement, and such arrangements do not always show on the deed. If in doubt, check with the previous owner or the local utilities.

On a broader scale, many provinces have **laws** that may limit sunwing design. Because of the distance between McDiarmid's house and the building next door, provincial fire-code regulations prevented any windows in the east wall of the addition, much to McDiarmid's chagrin, since his orchids would have welcomed the morning light. Many provinces also have laws to preserve the integrity of heritage buildings and sensitive natural environments. When Gerald and Joan Donnelly added a sunwing to their central Ontario farmhouse, the design had to be approved by the Niagara Escarpment Commission, which regulates all new construction within a swath of protected land from Tobermory to Niagara Falls. Although their plans were accepted as presented, the application took months to be processed.

Even when there are no special restrictions, getting plans approved can take a long time. Mary Coyle waited three months for her building permit because the local southwestern Ontario inspector sent the plans to Toronto for approval: in his experience, a "greenhouse" was something that stood in the middle of the yard, and he was not sure how to assess one connected to the house. Many municipalities still categorize greenhouses as farm buildings and have no provisions in their codes for the attached variety. Sometimes a simple name change can facilitate the process. When Eric Darwin first made his application to tear down the existing porch and build a sunwing, officials demanded a "design control" process that meant a six-month delay. So Darwin changed his proposal from "demolish existing building and build new" to "repair existing building by replacing walls, floors, windows, roof, services, etc.," and got his permit right away.

Architects and designers will likely be well versed in local, provincial and federal restrictions, but homeowners intent on doing for themselves should contact the local building inspector early in the process to find out what restrictions apply and how long it will take to unravel the red tape.

Aside from the legalities, every piece of property will undoubtedly present some practical limitations to sunwing location. Does a south-facing addition imply any major reorganization of the property, such as moving a driveway? Are there hydro and telephone lines overhead or underfoot? Are there sources of noise or visual pollution that should be avoided? Some parts of the property may simply make poor sunwing sites, like the corner across from the ball diamond. After the Meisners made a solar survey of their Nova Scotia home, they thought they were all set: the east-west axis provided a long wall facing the south. However, when they discovered that the septic system commandeered half the yard, they had to settle for a sunwing offset from the southwest corner of the house. Even so, it more than meets their expectations, providing "unlimited supplies of peppers, tomatoes, parsley, lettuce and herbs — and a pleasure that is difficult to describe."

When all local, provincial and federal regulations have been researched, mark on the property sketch (see page 31) those that might affect the addition — lot lines, road allowances, easements, Note any practical impediments to construction, as well as any special features that the sunwing should take advantage of, such as a lakeside view. Through this elimination process, the homeowner can reject sunwing locations that will cause problems and disappointment, leaving only the most promising sites for a passive solar addition.

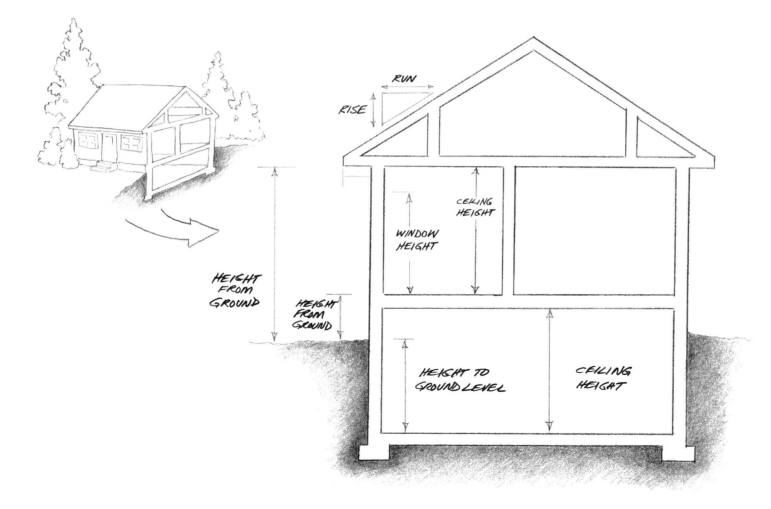

CROSS-SECTION
A few minutes with a measuring tape and sketchpad will ensure that the sunwing is more than just an architectural afterthought. To make the sunwing an attractive, functional part of the house, draw a cross-section as if the house were sliced from rooftop to footings at the point in the wall where the sunwing will be attached. Fill in the dimensions, as shown, then experiment with sunwing shapes and sizes.

TOP-LEVEL CONNECTIONS
Attaching the sunwing roof at different locations on the house can dramatically alter the living space. Extend the existing roofline, add a shed roof to the house wall, or design a two-storey space with the new roof affixed above the old.

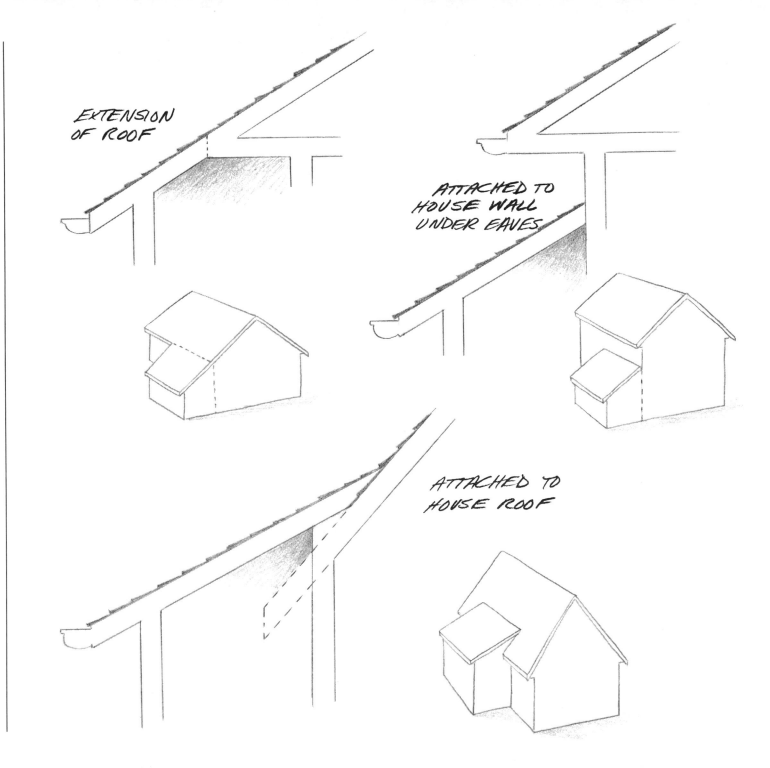

EXTENSION OF ROOF

ATTACHED TO HOUSE WALL UNDER EAVES

ATTACHED TO HOUSE ROOF

THE HOUSE

When the Donnellys decided to build a passive solar addition, they did not want it to look like an afterthought. "It mustn't offend the eye," they told their architect. They need not have worried. The finished sunwing is such a logical extension of their farmhouse that despite its solar design, it is hard to distinguish between new and old construction. The integration is more than skin deep: the rooms inside the house open so naturally into the sunwing that the Donnellys are drawn into the sun-filled space every day of the year.

Although such external factors as sun and site may eliminate many potential sunwing locations, the final choice is a matter of personal preference. Homeowners have to examine their house plan and their priorities so that the sunwing is attached where it will get the most use and where it can best fulfill its assigned functions.

To select the actual sunwing site, draw a floor plan of each level of the house, sketching it roughly to the same scale as the property plan so that the two can be overlaid. Pencil in the functions of each room and the location of doors and windows. Now compare the floor plan to the property plan, and pinpoint those areas not barred by the solar and site analyses. Look at each of the possible locations in turn, and imagine what the house would be like with a sunwing extended there. Would it fit in with the current use of the adjacent rooms? Consider which rooms would benefit most from the added light and heat, or whose privacy would be invaded by a transparent wall next door. Try to predict how the new space will alter the traffic patterns in the house, particularly if the addition has an outside entrance. Eric Darwin failed to realize until after

This two-level sunwing opens both the main floor and basement of the house to the sun. The upper floor, supported by posts and a beam, serves as a balcony that overlooks the sloped glazing and growing bed below. Max and Edith van den Berg, Winnipeg, Manitoba

he built his sunwing that it created a 15-foot barrier between the kitchen and the garden: he had to cut another door in the kitchen to restore direct access to the yard.

Use tracing paper laid over the floor plans to experiment with extending the house at different locations and different levels. A sunwing need not extend from the main floor. It can just as easily (and more cheaply) be built on a second or third storey, over a garage or bay window if the structure can bear the weight or be reinforced to support the new construction. A sunwing can also replace an existing part of the house. One Toronto couple razed their garage and on the same foundation built a new kitchen and sunwing/dining area, with a carport on the north side.

In fact, a sunwing may not be a wing at all. It may share two or even three walls with the main house, creating a semi-enclosed passive solar space. In this case, adding a sunwing can actually improve the shape of an existing house. In Canada's cold climate, complex house shapes, such as T, U or L, increase heat loss because of their greater outside surface area. The square is the most thermally efficient shape, and the best for solar gain is an elongated rectangle with an east-west axis. The sunwing may reshape the house closer to one of these ideals. As part of the overall design, consider changing a thermally inefficient north- or west-facing entrance to a more protected location on the east or south. When Harry Daemen designed a sunwing for his Cape Breton Island house, he used it to square off an L section at the front and to enclose an airlock entry and bootroom that cut down draughts into the house.

By looking at the house plan, the homeowner can decide exactly where to extend the sunwing. But to fully integrate the addition, one needs to see the house in cross-section. Draw a view of the house as if it were sliced through from roof to footings at the point in the wall where the sunwing will be attached. This cross-section shows the

roofline of the house, the floor levels and the level of the finished grade outside. Fill in the dimensions shown on the diagram on page 35.

The sunwing design can be developed from the bottom up, by first establishing the depth or north-south dimension of the addition: this will determine where the outside foundation will be. If the entire space will be used to grow plants and there are no artificial grow lights, winter sunlight will have to penetrate right to the back wall, so the sunwing should be no more than 10 feet deep. A solarium-sunwing with plants clustered around the south windows can be somewhat deeper, though generally no more than 15 feet, since long and narrow is the most solar-effective shape.

With the outside dimension established, the south wall can be sketched in. Commercial prefab greenhouses often show glazing curved

continuously from foundation to where it attaches to the house. Although in such cases, the distinction between roof and wall becomes blurred, the line that rises from the foundation can basically be either vertical or sloped. (For the pros and cons of that debate, turn to page 45.) If the glazing is sloped, it can rise directly from the foundation or from a low vertical kneewall and extend upward to meet a solid roof that attaches to the common wall. Depending on the level of the ground outside, a kneewall may be needed in snowbelt areas to raise the slope above winter accumulations. A kneewall will also provide extra headroom above the growing beds or space for tall plants to extend to their full height. Elizabeth Cran intended her greenhouse-sunwing to have a 2-foot kneewall, but owing to a misunderstanding with her contractor, the top of the kneewall became floor level. "This is somewhat inconvenient for plants," she says, "but I'm going to try vine plants like pumpkins and let them sprawl into the angle."

The top of the sketch connects the south wall of the sunwing back to the house. If the roof ridge of the house runs east-west, the slope of the house roof may be gentle enough to be simply extended down to meet the south wall. If the house roof is very high, the sunwing roof can be tucked under the eaves; or if the slope is too low, the sunwing roof may be attached to the top of the house roof. If the house ridge runs north-south, the sunwing roof can be attached to the wall or extended to the peak of the house.

When the roof and south wall are sketched in, only the floor needs to be added to completely define the interior space. The cross-sectional house plan shows the outdoor grade level and the house floor levels. Although these set

As well as allowing homeowners to open the second floor of their house to the sunwing, two-storey solar additions offer such design possibilities as vaulted ceilings, increased glazing and better heat collection. Gerald and Joan Donnelly, Albion Hills, Ontario

Gail Harvey

FLOOR-LEVEL
CONNECTIONS
Once the width and roofline
are set, experiment with
different floor levels. A
sunken floor gives more
headroom and brings the
outside landscape closer to
eye level, while a sunwing
with direct access from the
basement brings both
sunshine and plants to two
levels of the house.

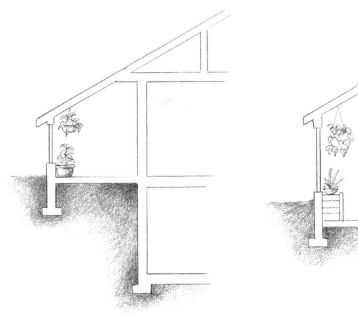

SAME LEVEL AS
MAIN FLOOR

A FEW STEPS
DOWN

BASEMENT LEVEL
ENTRY

points must be considered, the sunwing floor can float anywhere between, in an exciting array of options. It can be at the same level as the house's main floor, creating a smooth transition between indoors and outdoors. If the sunwing floor is lower than the adjacent room, the design must include space for steps, but the split level provides an interesting perspective for people looking down from the house into the sunny, plant-filled room below. If the grade level is at the house floor level, it can be very pleasant to step down into the sunwing and have growing beds at the same level as the grass and trees beyond. The final decision on floor level depends partly on entrance requirements – whether there will be doors to the main floor, the basement and outdoors – but it is also

linked to whether or not the homeowner wants a basement, cold cellar or crawl space underneath and how high the ceiling is to be.

Although energy-conscious architects used to frown on high-volume spaces, the trend in solarium-sunwings seems to be toward two-storey structures. The extra height adds drama to the exterior lines of the house and creates an exciting interior space that can accommodate large plants, even trees. "I've always felt that this was the only way to go if at all possible," says Michael Kerfoot of Sunergy Systems Ltd. in Carstairs, Alberta. "Almost all my new designs are two-storey. Thermally and psychologically, it is much preferred."

The increased glazing height invites winter sunshine to penetrate deeper, so

the space can be wider without sacrificing light. The extra expanse of glazing, however, will also cause greater heat loss in winter and overheating in spring and fall, but the extremes will not cause as much of a problem because of the large volume of air being heated and cooled. In very tall spaces, the stratification that naturally occurs as warm air rises can be used to advantage: the lower floor stays cool, while the heat near the peak is either vented outdoors, distributed to the main house or recirculated through thermal storage mass.

Having looked at the sunwing in cross-section, draw up a plan of the new space to determine the approximate size and layout. The New York State Energy Research and Development Authority's

David Quiring

Cross-sections and plan views do not give a very realistic impression of what the house will look like with a passive solar addition. To visualize the effect of a finished sunwing, sketch the addition on tracing paper laid over an enlarged photograph of the house.

solar greenhouse survey yielded some interesting comments about sunwing size. Most of the survey respondents had additions that were 180 square feet or less. Almost a third said that the usefulness of their additions could be improved by making them larger, and when size and satisfaction were compared, the homeowners with large sunwings (over 280 square feet) were most satisfied with their investment. Although cost will likely be a limiting factor, it is wise to anticipate needs as precisely as possible at the outset so that, once built, the sunwing proves large enough. Cutting costs by cutting square

footage may prove to be false economy in the long run if an addition to the addition becomes necessary.

In a greenhouse-sunwing, the floor area is determined by the number of plants the homeowner wants to grow. Most of the room is devoted to growing beds, which are usually 2 to 3 feet wide and arranged in rows in front of the glazing and north wall or in a peninsular design. According to Quebec energy consultant Ron Alward, co-author of *Low-Cost Passive Solar Greenhouses*, "With intensive growing practices, 60 square feet of growing-bed area is enough to supply one person's winter

greens. But that assumes some gardening skill and certainly won't provide room for vegetables like tomatoes. About half that area should be adequate if the person just wants to supplement the family's vegetable needs over the winter." A greenhouse-sunwing should also include a potting area and, if the outside door will be used in winter, an airlock entry.

To get a clear idea of room sizes, make a plan of the space with pencil, paper and scissors. Figure out the dimensions of furniture, growing beds, planting shelves, et cetera, that will go into the sunwing, and make scaled cutouts.

Move the cutouts around on graph paper, making sure to leave traffic areas between furniture and plants, and account for the space needed to open doors and windows. This should produce a fairly accurate estimate of the required sunwing size.

Before finalizing dimensions, consider the structural aspects of the sunwing. Take a look at the wall to which it will be attached, and, if possible, adjust the east-west dimensions of the sunwing so that its end walls line up with studs in the common wall. The south wall length should not be set until glazing is selected, for it should be built to accommodate available glazing sizes. Consider standard construction units in the planning: everything from ceramic tiles to ceiling joists come in standard widths and lengths, so building in appropriate increments will ensure a minimum of waste.

It is a good idea (and a lot of fun) to spend time experimenting with the plan and cross-sectional views of the proposed design. Try increasing the length of the roof, and watch what happens to the height of the south wall. Lower the floor a few feet to create a whole new space. The design process is circular, with a change in one dimension affecting two others. The next chapter goes through each design variable in detail, explaining how it will affect the performance of the space. Keep these sketches close at hand, and continue to circulate ideas, projecting them in plan and cross-section as the best possible sunwing evolves.

If you find it hard to appreciate the three-dimensionality of a design from section and plan drawings, try using a felt pen and some enlarged photographs of the existing house to visualize the sunwing. Put a large sheet of tracing paper over the photograph, and with the felt pen, draw the key lines of the house, including the roofline, gutters, windows and entrances. Then put a clean sheet of tracing paper over the line drawing, and add the sunwing. To get it the right size, use something in the photograph as a guide. For example, if the window is 4 feet wide and the addition is 16 feet, measure the window in the photograph and multiply by 4. Usually, the lines of the addition will be parallel to or an extension of existing lines of the house. When the basic shape of the sunwing is roughed in, add windows, skylights, landscaping, even clouds in the sky, to help visualize the finished sunwing before it is built. Draw several variations of each idea: tracing paper is a lot cheaper than lumber.

Using an actual photograph of the house will also help you to design an addition that complements the style of the existing house, rather than something that looks like it has been pasted on in the dark. With a little extra planning, a sunwing can be aesthetically as well as functionally integrated into the house. Sometimes it is the small details that create visual links between new and old construction, things like recreating the style of soffits and trim or using windows and doors with similar proportions to those in the rest of the house.

This process is not a job to be polished off on a weekend: it may take weeks, months or even years for those who enjoy the process of change as much as the change itself. The words to keep firmly in mind are "what if . . . ." They will stimulate some of the craziest ideas and will likewise rule out a lot of absurd inspirations that seemed truly brilliant at first thought. Darwin went through at least 10 variations of his design – the most complex was a multistorey version with balconies and spiral staircases. "At that point, we said, 'Whoa: we don't want a house that is an accessory to an addition!' So we reined ourselves in, looked at what we really wanted and ended up with something we are very pleased with, that looks like it belongs on the house."

Jim Merrithew

Although the design for this house was examined by professionals, the owners first spent more than a year devising and rejecting almost a dozen versions of their original concept. Eric Darwin and Frances Dubois, Ottawa, Ontario

4 Solar Sinews

Elements of solar design

'The sun never repents of the good he does, Nor does he ever demand a recompense.'
— Benjamin Franklin

So far, the sunwing has taken shape according to a series of givens — the sun, the site and the house itself. Now comes the exciting part: choosing the design elements and integrating them into a unique solar space.

There are few absolutes that apply. Because all the elements are interdependent, each choice creates a ripple effect. For instance, if sloped glazing is chosen to blend in with a Cape Cod roof, some form of storage mass will be needed to counteract the thermal swings and shade controls for the summer months. The final design evolves out of a round robin of such decisions. To be successful, the process must have as its centre a firm grasp of what a sunwing is to do.

"One of the hardest things in self-design is figuring out priorities," says Don Roscoe, who routinely guides dozens of homeowners through Solar Nova Scotia's greenhouse course. "What you have to do is make the things that are most important to you work the best, and be satisfied if the least important things just work okay."

Many of the decisions that once were simply a matter of personal taste — what colour to paint the walls, where to place the windows — must now take into account the effect on thermal performance. Given the Canadian climate, there is a tendency to think only in terms of the heating season, but unless the space will be closed off completely at certain times of the year, thermal performance has to be projected through all seasons. For instance, west glazing adds extra light in winter, but in summer, it becomes a liability, causing afternoon overheating.

Imagination can sometimes turn such liabilities into design advantages. When fire codes prohibited windows in the east wall of Garnet McDiarmid's greenhouse-sunwing, by building the wall of brick, he turned it into a useful thermal storage mass to temper the overheating caused by the substituted southwest glazing. New England solar designer and inventor Norman Saunders calls this "double-duty integration": carefully matching components to work together cooperatively, each making up for the limitations of the others.

Furthermore, each component should be designed to do more than one job. American solar pioneer Steve Baer always tries to kill three birds with each design stone. South-facing windows, for example, admit light and view and collect solar radiation, even though they have an unimpressive R value. Keep this multipronged approach in mind throughout the chapter: it will not only streamline the design and improve efficiency but will probably save money.

GLAZING

Early passive solar homes bared broad flanks of angled glazing to the sun, and today, that trademark of "solar chic" graces everything from suburban malls to muffler shops. Far from being the harbingers of the future, however, such overglazed structures are architectural dinosaurs. Freezing in winter, roasting in summer, they require massive infusions of heating and cooling energy dollars to keep them comfortable year-round.

In a cold climate like Canada's, glazing has a dual nature: although it controls the light and heat energy that a building collects, it is also the primary source of heat loss when the sun is not shining. If a passive solar addition includes enough glazing to heat the room on the coldest, shortest day of the year, it will be overglazed for the remaining 364 days, collecting — and losing — more heat than is necessary.

"Sunspaces by definition are

Jurgen Mohr

Although it is prone to overheating in summer and excessive heat loss in winter, sloped glazing provides high levels of light, especially in spring and fall, and is ideal for a plant-growing space. Peter and Inez Platenius, Verona, Ontario

The high glazing in the east part of this two-section greenhouse-sunwing (previous page) maximizes light in the two-level growing space. The west section was kept low to retain the view from the house windows. Ross and Vivien Rogers, North Bay, Ontario

overglazed,'' says Ottawa designer Chris Jalkotzy, ''but some are more so than others. I have a hard time convincing clients that the room will still be bright and sunny even if the south-facing window area is cut back a bit. Luckily, I have designed enough additions that I can show people that they don't have to glaze the entire east, west and south walls to have a sunny space.''

So how much glazing is enough? Solar manuals usually calculate the area of south glazing by figuring out how much solar gain is needed to heat the building. However, since passive solar additions are not designed as solar furnaces, the amount of glazing should not be decided solely on the basis of how much heat it will collect. David Bergmark echoes the sentiments of many designers when he says, ''There is no such thing as an optimal percentage of glazing for these generalized spaces. The important issue is that a unit area of glazing has a certain potential for heat loss and heat gain.''

Homeowners should therefore include the glazing they want, giving due consideration to the effects of slope and position. The actual heat loss and gain will depend on orientation, shading, climate and the glazing material itself. Working out the figures is not as important as incorporating the appropriate thermal controls for the proposed area of glazing: heat-loss controls to improve winter comfort, shading to prevent unwanted heat gain, and storage and air-handling systems to take the edge off the thermal swings.

''People seek natural daylighting, viewing, psychological expansiveness,'' explains Michael Kerfoot. ''Meet these needs first. If glazing is excessive, as it usually is, then integrate strategies for tempering. Don't design the other way around.''

Martin Foss used his pocketbook to

determine the amount of glazing he could have in his Ottawa area sunwing. "The plan (in the mind, of course) called for an insulated concrete slab floor, 2-by-8 framing for walls and roof and as much glazing as we could afford." As haphazard as it sounds, the design works. "Sunny days are sublime!" enthuses Foss. "The sunspace heats up to 24 degrees C [75 degrees F] by 9:30 a.m. and eliminates the need to keep the two wood stoves operating in other parts of the house. The slab floor absorbs heat effectively, and stocking feet can be worn comfortably. In late fall and early spring, the sunspace really comes into its own, as the heat it generates is sufficient to warm the whole house." Foss expected the sunwing to overheat in summer, but the problem never materialized because he installed his glazing vertically.

Indeed, vertical versus sloped glazing is one of the major debates in sunwing design. In midwinter, because of the low angle of the sun, vertical and sloped south walls collect about the same amount of solar heat energy. In fact, vertical glazing may collect slightly more because it benefits from reflected radiation bouncing up off the snow. As the sun rises in the sky, the sloped glazing collects more and more solar radiation, although as the days themselves are getting warmer, the sunwing needs the heat less. For instance, in Edmonton at Christmas, the average daily solar radiation striking a south-facing vertical wall is only 14 watt-hours per square yard less than what strikes the same wall sloped to 60 degrees. By Easter, the sloped wall receives 1,300 watt-hours per square yard more radiation than the vertical surface. Although the warmth may be appreciated at the shoulders of the heating season, in the late spring and

---

The Great Glazing Debate

VERTICAL

- easy to install
- easy to seal
- cheaper (not tempered, less labour to install)
- good winter solar gain
- little summer solar gain
- less spring/fall gain — may need roof glazing to perform well as growing space in late spring
- more energy-efficient
- less prone to overheating — more useful in spring/summer/fall

SLOPED

- harder to install
- harder to seal
- more expensive — glass should be tempered
- good winter solar gain
- good spring/fall solar gain — best for plants
- prone to overheating — shade controls and vents required
- increased visibility
- dramatic interior space
- less volume of air to heat
- more heat loss through overhead slope

---

fall, sloped glazing will cause serious overheating. When Gary and Sandy Hodson added a sunwing to their central Nova Scotia home, they sloped the glazing 45 degrees to match the roof angle. Even in mid-March, three west wall vents have to be opened to flush unwanted heat outdoors.

"Where there is overheating, there is bound to be overcooling," warns Brian Marshall. "People don't realize it because the backup heat makes sure the space is still comfortable, but the heat losses through sloped glazing are incredible."

Because warm air rises, sloped glazing offers the least thermal resistance where it is needed most. Instead of an R30 roof, there is only R2 glazing between the heated sunwing and the cold outdoors. Due to its angle, sloped glass is virtually impossible to fit with effective inside curtains. Glazing in this position is also

difficult to clean and is vulnerable to damage from falling objects such as icicles, branches and hail.

Sloped glazing requires more care in installation. It is prone to leaking, especially if runoff is dammed by cross-members, and when double insulating glass units are used, the bottoms of both panes must be supported to prevent the seal from shearing. Instead of sliding off, snow tends to sit on a low slope, creating ice dams as it melts. The extra weight may require a load-bearing beam where the slope meets the roof. Because it requires time-consuming detailing and sealing, construction labour costs may be higher, as will material costs, since the sloped glass should be tempered.

A low slope, such as that in a glazed roof, creates more problems than a south wall that is slightly tilted, and in some cases, a gentle slope has distinct

Robert Tinker

The glazing on this owner-built sunwing is tilted at 80 degrees. Though it increases midwinter insolation, even this gentle slope causes some overheating in spring and fall. Max and Edith van den Berg, Winnipeg, Manitoba

advantages over vertical glazing. A sunwing with angled glazing benefits more from general diffuse sky radiation, and direct sunshine can penetrate more deeply into the room. This is important in a greenhouse-sunwing, where growing beds may extend more than 2 feet inside the south wall. (Even with sloped glazing, however, light levels in December and January will probably be too low for active growth.) Aesthetically, a slope is dramatic, adding sweeping lines to the otherwise boxy appearance of a house. It also gives people a greater visual connection to the outdoors. After Michael Kerfoot explained all the disadvantages of sloped glass to one of his clients, she replied, "Yes, we understand, but we want to be able to look up and see the stars!"

If the proposed design includes vertical glazing and a band of overhead glazing for plant light, sloping the entire south wall may actually be more practical. It will be easier to build because a tricky glazing joint is eliminated, and cheaper because there is less side wall area to construct. With less total glazed area, heat loss will be cut, and there will be a smaller volume of air to heat.

Homeowners who opt for sloped glass must build in adequate shading, ventilation and storage systems to counteract excessive solar heat gain. The glazing should not extend all the way to the common wall but should be attached to an insulated solid roof to minimize heat loss and to shade the interior in summer. For good light, a slope of about 65 degrees (latitude plus 20 degrees) is appropriate for most of Canada. Areas subject to a lot of cloud and fog can tolerate a slightly flatter angle, while very clear sunny sites should have glazing more steeply angled.

Nevertheless, the trend among Canadian designers is toward vertical south walls, especially if the sunwing is used primarily as a solarium. Vertical glazing has comparable heat gain during the coldest part of the year, is easier and therefore less expensive to install, can be fitted with movable insulation to cut night heat loss, and does not overheat, since outside of the heating season, the sun rises high enough that it does not shine directly into the space. Although it does not provide enough well-distributed light for efficient midwinter plant growth (especially vegetables), there is plenty of light for starting seedlings and growing most houseplants. They may tend to lean toward the single light source, but this can be corrected by rotating the plants.

One way to increase light levels without resorting to sloped glazing is to increase the height of the south vertical wall relative to floor depth. High glazing will let light penetrate more deeply into the sunwing, even in the spring and fall when the sun rises higher in the sky. This extra height creates its own compromises — extra expense in

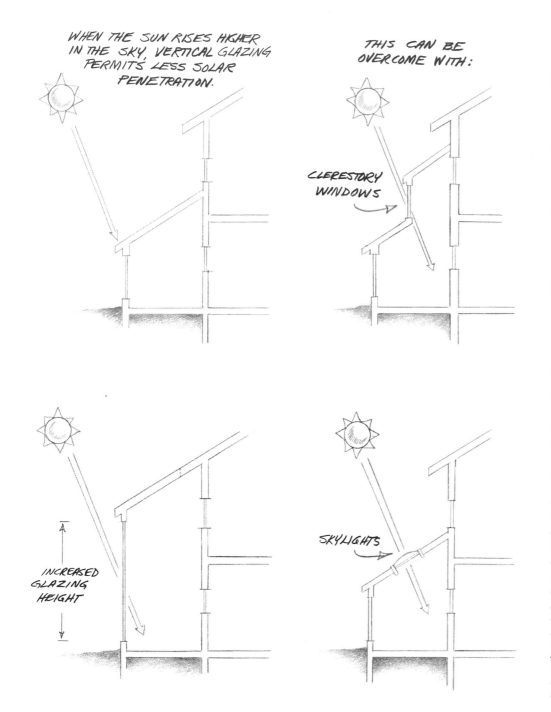

WHEN THE SUN RISES HIGHER IN THE SKY, VERTICAL GLAZING PERMITS LESS SOLAR PENETRATION.

THIS CAN BE OVERCOME WITH:

CLERESTORY WINDOWS

INCREASED GLAZING HEIGHT

SKYLIGHTS

construction, additional heat loss because of increased glazed area and stratification, and potential summer overheating. That extra height does not necessarily entail a tall south wall, however. Eric Darwin brought light to the centre of his sunwing with clerestory windows. The north half of the sunwing roof is raised 4 feet above the south half so that south-facing windows in this upper section admit extra sunshine as well as venting summer heat build-up.

Another way to enjoy the advantages of sloped glazing without incurring all of its drawbacks is to install skylights in the solid roof. Because they are thermally inefficient, they should be used discriminately and detailed carefully to prevent leaking, but if installed properly, they can provide some of the overhead light so necessary for plant growth and daylighting. Unlike a solid bank of overhead glazing, skylights do not need shading devices, because there is enough shaded floor area to which plants and people can move when direct sunlight is too intense.

Skylights can also be incorporated into the design to improve daylighting in a very wide sunwing: a narrow strip of skylights close to the north edge of the roof is most effective in wide sunwings, because it washes the north wall in light that is reflected back into the room. Skylights can also be used to send a shaft of sunlight into the main house. Ironically, a sunwing can actually darken the interior of a house if too many windows are blocked by the new construction. If the skylights are placed directly opposite existing windows, those rooms will not be cut off from the view and light they now enjoy. When John Hix designed a sunwing for the Donnellys, Joan's mother dubbed him Mr. W.B. (Mr. Window Blocker) because she thought her second-floor bedroom

LIGHT WITHOUT HEAT LOSS
The relatively low light levels that are the major drawback of vertical glazing can be overcome by designing in skylights, clerestories or very high windows, which are generally more energy-efficient, cheaper and less problematic than sloped glazing.

French doors help to counter the tendency of this sunwing to darken the interior of the house. On sunny winter days, the doors are opened to provide warmth as well as light. Gerald and Joan Donnelly, Albion Hills, Ontario

growth. Most vegetables prefer to be outdoors in full sun all day and, grown indoors, will take as much direct sunlight as they can get. Yields will be minimal or plants will stop growing altogether if light levels are too low. The actual amount of light required varies with the species: for pollination alone, tomatoes require four hours of direct sunlight a day.

When Joan and Gerald Donnelly built their sunwing, they insisted on a west window because, as farmers, they were loath to lose their view of the swan pond, where their prize waterfowl swam. Aside from retaining a necessary or cherished view, side wall glazing may also be necessary to provide more even light and to reduce glare. If the sunwing is deeper than it is wide, exclusively south glazing will create an unpleasant, cavelike effect.

If side wall glazing is incorporated into the design, an east window is preferable to west: it admits early morning light for the plants without overheating the room. The glazed area should be minimal and fitted with movable insulation, preferably high R-value shutters that can be tightly weatherstripped and installed permanently during the coldest months of the year. If seasonal east-west windows are made operable, they can double as vents, opening for breezes, light and view from spring to fall, then be blocked off during the winter to reduce heat loss.

The common wall shared by the house and the sunwing can also be partially glazed. Double-glazed windows and doors (patio or French) can inject light into the house, contribute to both warm air distribution and ventilation and provide a visual and functional transition between the house and the sunwing. Because the temperature differential between the sunwing and

window would be covered by the new roof. Instead, a skylight directly opposite the window preserves the traditional view of the maple bush across the road, as well as providing an overview of the lush green plants in the sunwing below.

East-west walls should have as little glazing as possible, since they lose heat in winter and gain too much in summer. If a plant-producing sunwing is at least twice as long as it is wide and faces solar south, east-west glazing will not be necessary. However, if the addition is more than 15 degrees off south, it will need windows in the side wall (or a low slope) to provide enough light for active

Gail Harvey

house is less than that between house and outside, there will be little heat loss through common-wall glazing and possibly some radiant heat gain.

A sunwing may also improve the effectiveness of existing south windows. In one Toronto home, louvred blinds were always drawn in front of the kitchen patio door because the intense outdoor sunlight was too hot and bright in summer. When a sunwing was added to that side of the house, the blinds were retired and the darkened kitchen came to life. The patio door still admits direct heat gain in winter, but it is shaded from the summer sun so that sunlight can filter in without overheating one of the most used rooms in the house.

THERMAL STORAGE MASS

On a sunny day, temperatures in a highly glazed room will quickly soar beyond human or plant comfort levels. By the same token, because the glazing has only minimal heat resistance, when the sun goes down, the room loses heat rapidly and temperatures plummet: the mercury in a sunwing can fluctuate from the ambient night temperature to as much as 150 degrees F on a clear winter day. Plants will not survive such conditions, since they generally need a minimum of 55 degrees and a maximum of 85 degrees for healthy growth. Although people are not as sensitive and demanding as plants, wildly fluctuating extremes will obviously limit the usefulness of an addition. This leaves two options: excess heat can be pumped directly into the main house during the day, and backup heat can be used to maintain acceptable nighttime lows in the sunwing; or the sunwing can include thermal mass to moderate the extremes.

Any discussion of thermal storage mass is fraught with controversy, particularly as it applies to small spaces

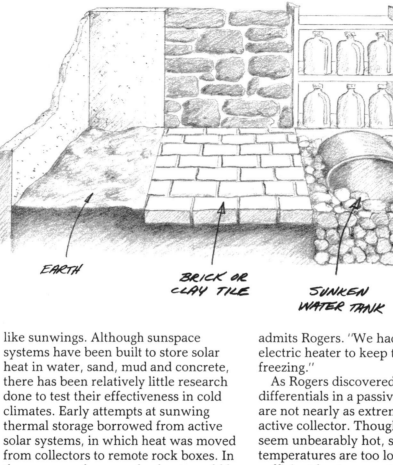

CONCRETE   ROCK WALL   WATER CONTAINERS

CEMENT BLOCK

EARTH

BRICK OR CLAY TILE

SUNKEN WATER TANK

ROCK

THERMAL STORAGE MATERIALS
Water, rock and concrete can moderate the extreme temperature swings typical of a highly glazed room, but they are more effective at blunting the highs than at tempering the lows.

like sunwings. Although sunspace systems have been built to store solar heat in water, sand, mud and concrete, there has been relatively little research done to test their effectiveness in cold climates. Early attempts at sunwing thermal storage borrowed from active solar systems, in which heat was moved from collectors to remote rock boxes. In theory, enough sunny-day heat would be stored to compensate for nighttime and cloudy-day losses so that the storage system would become, in effect, the backup heat itself. Ross Rogers installed a typical system in his 12-foot-square greenhouse-sunwing in North Bay, Ontario. A reversible air-handling system collects heat from the sunwing ceiling, stores it in a rock bed, then redistributes it through perimeter registers when sunwing temperatures fall. "It never really worked properly,"

admits Rogers. "We had to use an electric heater to keep the place above freezing."

As Rogers discovered, the temperature differentials in a passive solar addition are not nearly as extreme as those in an active collector. Though the room may seem unbearably hot, sunwing temperatures are too low to transfer sufficient heat energy to storage to keep the space consistently comfortable after the sun goes down. Considering the heat energy available, the mechanisms to extract it are extremely costly.

Although impractical and unreliable as a demand heating system, thermal storage can effectively blunt temperature extremes in a sunwing. In experiments at the National Research Council, passive thermal mass shaved 30 F degrees off the 100-degree range in a test sunspace. However, it was much more effective at the high end of the scale, lowering the peak temperature by 22 degrees while raising the night-time low by only 8 degrees. Since these few degrees will not be enough to keep most sunwings from freezing and there are much cheaper ways of dumping excess heat, the homeowner is strongly advised to balance carefully the cost of thermal mass against other systems that can do the same job.

"Remote active storage systems are fairly useless in additions," says David Bergmark. "When you stop and calculate how much it costs to provide that bit of extra heat energy, you could pay for the backup with the interest on the investment."

The same argument can be applied to **phase-change materials**, a relatively new form of concentrated storage. The principle is simple: when an ice cube changes to its water phase, it absorbs heat; when it changes back to its ice phase, it releases heat. Rods containing phase-change salts are compact and can store large amounts of heat at a uniform temperature, but the technology is still unreliable and very expensive. George Dewar bought 20 tubes of phase-change salts for his St. Catharines sunwing, hoping to provide dependable winter lows for his plants. "To be blunt, it was a complete waste of $800. There just isn't enough sun when we need the heat, or at least not a steady enough concentration of it. For me, they were absolutely not worth it."

Water-filled drums are a cheaper form of concentrated storage, and one that has become virtually a trademark of attached solar greenhouses. Although water can store more heat than most other media, it is not as effective at absorbing and reradiating solar energy as many have assumed. It heats up more slowly than rock, brick or masonry, and instead of warming up to a uniform temperature, the stored heat tends to stratify. Consequently, the stored heat does not radiate evenly back into the space because it is concentrated in one location — at the top of the containers. To compensate, large volumes of water have to be positioned so the bottom is exposed to direct sunlight and the top is insulated and shaded. Even so, in Wayne Wilkinson's experience, "they are a waste of time. They never warm up: the surface area is just too small to heat up the volume of water."

Laying the drums on their sides in front of a glazed kneewall counters stratification and improves performance somewhat. In that position, their functional nature can also be masked by a couch or planting bench. A better strategy is to divide the 2 to 4 Imperial gallons of water required for each square foot of south glazing into small containers and distribute them evenly throughout the sunwing. If they are spread around the floor, however, they leave little room for growing beds or furniture. And if stacked against the wall, both the floor and the containers must be able to stand the weight of the water (8.4 pounds per gallon).

Besides being heavy, water poses the problem of potential leakage. Damage can be prevented beforehand by building a drainage pan under the water-storage area that can hold the contents of at least one container, with a drain to an outside sewer or catch basin. If there is no guaranteed backup in the space, the water needs additives to prevent freezing, and in metal containers, it may need corrosion inhibitors.

"Luckily, we came along after rock bins and water walls," says Brian Marshall. "The best storage is the simplest — built right into the structure itself." Marshall is referring to **integrated thermal storage**, in which the walls, floor and interior finishes of the sunwing are made of materials that absorb solar heat directly, then reradiate it into the room passively, using no

The 1-gallon water containers concealed under this potting table are used to fine-tune the substantial heat storage mass in the foundation and the flagstone floor. Harry Daemen, Whycocomagh, Nova Scotia

Harry Daemen

mechanical distribution or collection systems. This approach is relatively inexpensive, unobtrusive and requires no maintenance.

One of the most popular forms of integrated thermal mass is a concrete floor slab. Concrete can store 30 BTUs of heat per cubic foot per 1 F degree rise in temperature, and as a floor, it is evenly distributed throughout the sunwing, making it both thermally efficient and structurally uncomplicated. Brick stores almost as much heat as concrete and can be laid over a wood subfloor or used to face the north wall of the sunwing. (If the outside wall of the house where the sunwing will be attached is already bricked, it can be used as thermal mass, provided there is a thermal break between the enclosed brick and the outside so warm air is not wicked out.)

Such concrete floors and brick walls provide useful but limited thermal storage mass. Heat penetrates masonry slowly, and the heat collected at the surface tends to be released to the air instead of being absorbed deeper into the material. "Only the first inch counts," says Michael Glover. "There is no point in paying for extra thickness, since anything beyond a few inches is useless from a thermal point of view."

Furthermore, integrated mass is indiscriminate in the heat it absorbs, storing backup heat as readily as free solar gain. It may therefore cost more than it saves in a heated sunwing. For instance, a sunwing may be used as a breakfast room, with electric heat to take off the morning chill. The integrated storage mass, which has cooled down overnight, is charged with expensive electric heat and is then unable to absorb the free solar gain that peaks later in the day. As a result, integrated storage is most useful in a buffered sunwing with no supplementary heat source or in one

Though this 14-foot-high common wall is thermally separated from the rest of the house exterior, the bricks do not provide enough heat storage to keep temperatures in the sunwing above freezing at night. Eric Darwin and Frances Dubois, Ottawa, Ontario

Jim Merrithew

51

**HYBRID INTEGRATED
THERMAL STORAGE**
A low-speed fan draws solar-
heated air from the sunwing
peak and blows through the
concrete floor slab. This
helps prevent daytime
overheating and raises
nighttime temperatures
marginally as the stored
heat radiates passively back
into the room.

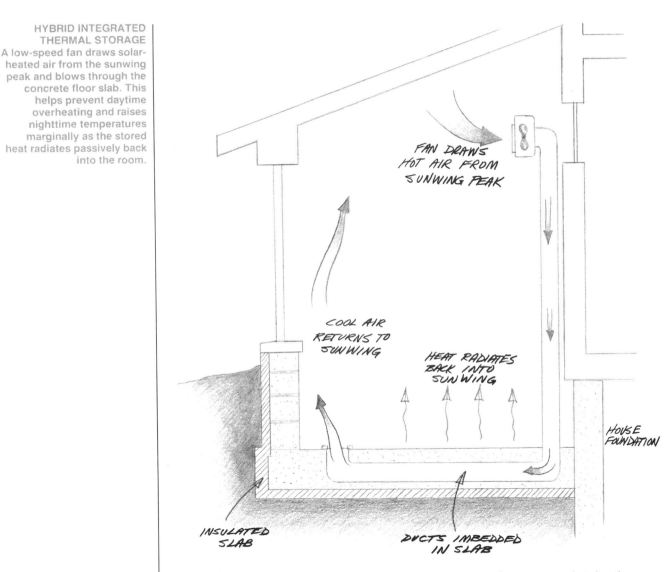

FAN DRAWS
HOT AIR FROM
SUNWING PEAK

COOL AIR
RETURNS TO
SUNWING

HEAT RADIATES
BACK INTO
SUNWING

HOUSE
FOUNDATION

INSULATED
SLAB

DUCTS IMBEDDED
IN SLAB

which is heated only in the evenings.

Integrated storage mass is most effective if exposed to direct sunlight, and the right finish can enhance its heat-absorbing abilities. A bare concrete slab absorbs about 60 percent of the solar radiation that strikes it; painted dull, flat black, absorption rises to 94 percent. Brown, grey and red also score high on the absorption scale. Rough texture increases surface area, so brick, tile or stone should be left unglazed or sealed with a nonreflective satin finish.

Even with a good absorptive surface, however, the floor is not the best place for passive thermal mass. "If there is any kind of kneewall at all, the floor won't get much direct radiation in winter, and the exposed surface area is usually limited by furniture and plants," says Kalev Ruberg, a solar researcher at the National Research Council. "Our studies showed that even directly illuminated bricks on the north wall can clamp temperature swings by only 3 Celsius degrees [5 Fahrenheit degrees] at the low end of the scale."

Unlike the floor, a wall is able to take advantage of the natural tendency of warm air to rise. However, the warmest air in the sunwing pools at the ceiling, and to take advantage of this, many designers now use a hybrid integrated storage system to make floor mass more effective: a low-speed fan draws hot air from the sunwing peak and moves it down to floor level, where it is absorbed and reradiated passively. The constant cycle of moving air is particularly beneficial for a growing space, since it reduces the problem of mildew and mould and keeps the plants supplied with fresh, carbon-dioxide-rich air.

John Hix combines air recirculation and minimal passive storage with an interesting floor design. Six-inch-diameter pipes are set against the northeast and northwest corners of his sunwings, with heat sensors at the top that turn on a fan when ceiling air temperatures reach a predetermined level. The fans pump hot air from the ceiling to the crawl space and across 6 inches of sand. The air then rises naturally back into the sunwing through a slotted floor constructed of slightly spaced 2 by 4s on end. Even this simple loop seems to temper extremes, at a relatively low cost.

Luc Muyldermans of Ayer's Cliff, Quebec, uses fan circulation to deliver heat directly to the plants. The peak air is blown through drainage pipes buried 2 feet below the surface of the growing-bed soil. "This soil-heating system is very favourable for growing," says Muyldermans, who harvests vegetables all winter from a completely unheated greenhouse-sunwing. "With heated soil, plants grow faster and are better able to resist low air temperatures."

Don Roscoe imbeds hot air ducts right in the concrete slab when it is poured. As the warm air moves through the metal ducts, its heat passes into the concrete, then radiates passively back up into the sunwing. According to Roscoe, the benefits are worth the cost of the ductwork and the thickened slab necessary to cover the pipes.

Heat can also be blown under the floor slab into a layer of sand, gravel or rocks. Unlike remote active storage, there is no attempt to control the retrieval of heat: it radiates passively back into the sunwing. In Michael Kerfoot's view, "if you have to deal with excess solar gain, this is the cheapest way to go, since the foundation walls and the slab are already there. The only cost is the rocks, pipes and fan, which is minimal."

Because the dynamics of storage mass are still not fully researched, the best guide is the budget. If mass can be incorporated at little or no expense and it will not be in a position to absorb paid-for heat, then it is a good way to blunt the temperature extremes of a sunwing, particularly in a mild climate, where a little boost will keep the sunwing above freezing. If, however, storage mass requires a considerable investment of money, time or energy, the best option is simply to vent excess solar gain into the house. In all but superinsulated houses, the sunwing heat will be welcome and will probably make enough of a contribution to home heating to balance the cost of maintaining a minimum low in the addition. Consult a solar professional before finalizing a thermal storage system, but keep in mind the advice of David Bergmark, who has had experience with active and passive

Roger Vernon

A concealed duct against the ceiling of this sunwing's north wall gathers hot air, which is then blown through 50 cubic yards of rocks stored under the tile-covered slab floor. Though it has no backup heat, temperatures in the addition have never gone below 55 degrees F. Greg and Linda Powell, Priddis, Alberta

thermal storage through rocks, water and mud: "The simplest always worked the best."

HEAT DISTRIBUTION

Even with thermal storage, on a sunny February day, a passive solar addition will probably collect more heat than can be absorbed. If heat is allowed to build up in the sunwing, plants will be damaged and heat loss will increase because of the differential between indoor and outdoor temperatures. Instead, during the winter, excess solar-heated air should be vented into the house, improving conditions on both sides of the common wall.

Generally, the moist, oxygen-rich air

produced by plants will be welcome in winter-parched houses, but if a greenhouse-sunwing is attached to a new house built to high insulation and air/vapour barrier standards, the excess humidity can be damaging. If windows in the house start to fog up and condensation appears on cool surfaces, heat exchange between the two spaces should be stopped.

The amount of heat there is to distribute, and thus the complexity of the system, will depend on how much solar gain the sunwing collects and absorbs and how the space is used. In a

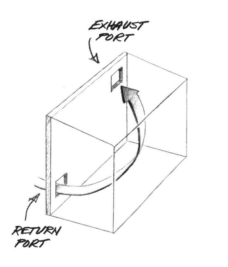

solarium-sunwing where collection is one of the priorities, the heat-distribution system can be an automatically controlled network of ducts and fans. In most small sunwings, however, "door control" is all that is required. When it is warmer in the sunwing than in the house, the connecting door is opened: warm air flows into the house through the top third of the doorway, and cool air

returns through the bottom. A double-hung window that can be opened both top and bottom would also work, though not as well, unless the window is very tall.

If the sunwing covers two storeys of the main house wall, the heated air should enter the house at the top of the second storey, with a cool air return port at the bottom of the first floor. It is important that there is a clear passage inside the house for the air to circulate back down to the first floor, completing the loop. In the Donnelly sunwing, warm air enters through upstairs

bedroom windows, which are situated close to doors that open onto a stairwell, at the bottom of which are the French doors that lead to the sunwing. This creates a very efficient airway.

If there are no appropriate common-wall openings or if no one will be home to open the doors and windows, special heat-distribution ports can be cut. To be effective, these wall openings must be correctly sized and positioned. There

should be openings near the top of the common wall to exhaust warm air into the house and openings near floor level to return cool air from the house. This creates a thermosiphon, a natural convective loop in which exiting warm air is constantly replaced by cooler return air.

Make the high exhaust openings one-third larger than the lower openings, which should return air from a cooler part of the house if possible. For instance, Elizabeth Cran of Tignish, Prince Edward Island, had heat-distribution ports cut in the common wall so that warm sunwing air rose into the bedroom and cool air returned from the basement below. The greater the height between exhaust and return ports, the more efficient the airflow will be; therefore, separate high and low ports vertically by at least 6 feet. They should not be directly above and below each other but offset to force the air to travel farther in its loop. Depending on the size of the sunwing and the layout of rooms adjacent to the common wall, more than one exhaust and return opening may be needed.

According to the *Seymour-Dorgan Report* published in 1985 by the National Research Council, for purely passive heat distribution, the total opening area should be no less than 2.5 percent of the sunwing glazing area. For instance, in a sunwing 10 feet by 16 feet with the south wall entirely glazed ($7 \times 14 = 98$ square feet), there should be no less than 3 square feet of heat-distribution ports. This is an absolute minimum, and most designers incorporate more. Don Roscoe suggests that high and low ports equal 6 to 8 percent of the floor area, or if windows alone are used, they should represent 30 to 45 percent of the common-wall area.

During the heating season, all

common-wall openings should be closed at night to prevent reverse thermosiphoning — warm house air flowing into the sunwing and cool sunwing air being drawn into the house. Ports should be fitted with back-draught dampers, which can be a flap of polyethylene film if the sunwing temperature is maintained above freezing or an insulated shutter if the sunwing is unheated. According to Solar Nova Scotia's calculations, if the port is screened, its size should be increased by half to compensate for the reduced airflow.

In sunwings with little thermal mass or a lot of sloped glazing, heat can build up faster than natural convection can carry it away. A portable fan set in the window or doorway to speed up distribution may be enough to prevent overheating. Indeed, many homeowners have found that fans are necessary to give passive airflow a push in the right direction. As Brian Marshall has discovered, "natural convection doesn't always flow the way you want it to." Without mechanical assistance, warm air may be more inclined to flow down the cool, sloped glazing than to rise through ports into the house.

Fans can also be used with ducts to transfer sunwing heat to rooms elsewhere in the house. A clean, dry, insulated basement is an ideal destination for solar-heated air, though if it is very humid and the basement is unheated, condensation may occur. If ducts travel through unheated spaces, they should be insulated.

Fan sizes are measured in cubic feet of air per minute (CFM). The size of the fan needed will depend on how much excess heat there is to distribute. As a general rule, the *Seymour-Dorgan Report* recommends that the fan should be capable of moving the entire volume of

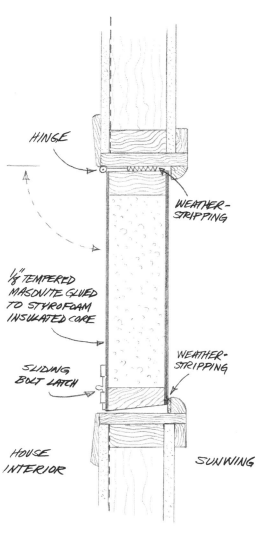

HINGE

WEATHER-STRIPPING

⅛" TEMPERED MASONITE GLUED TO STYROFOAM INSULATED CORE

SLIDING BOLT LATCH

WEATHER-STRIPPING

HOUSE INTERIOR

SUNWING

greenhouse air into the house in two to four minutes. To determine the correct size, calculate the volume of the sunwing, and divide by 2 if a large volume of excess heat is anticipated, or by 4 if the sunwing will contain some thermal mass. Dividing the capacity required among two or three smaller fans evenly spaced along the top of the common wall will be more effective in distributing heated air to the house than

a single large fan. Kitchen exhaust fans or room-to-room fans, available from wood-stove dealers, are appropriate. Consult a heating specialist to determine correct sizing for ducts if required. Remember that fans should always have back-draught check dampers.

Although they improve the efficiency of the sunwing as a collector, fans do have drawbacks. They can create unpleasant draughts, and the constant whirring of the blades is often an unwelcome distraction in a space designed as a sunny retreat. Noise levels vary widely among different products, so it is worthwhile to test several models before buying. Paying more for a quiet fan is money very well spent. The fan also represents a reliance on a machine, with the expense and maintenance that entails. In a greenhouse-sunwing, where heat control is critical, there should be an alternative way to flush heat out of the sunwing in case of power failures.

Both active and passive heat-distribution systems can have manual or automatic controls. For manual control, the homeowner has to be home to turn on the fan or open the window when the sunwing is warmer than the house. An indoor-outdoor thermometer, with the outdoor bulb attached in the sunwing, will indicate when there should be an air exchange between the two spaces. A system that relies on manual control will work like clockwork only if the operator is equally precise.

Automatic controls for fans and ports are more reliable and more expensive. The simplest form of automation is a **thermally activated control device** (TACD). A change in air temperature causes a liquid inside a piston to expand or contract, pushing the port open or closed. As one homeowner discovered, however, when he tried to have a paraffin piston open a heavy cedar door,

TACDs are limited to relatively small loads and work best if installed at the top of the sunspace where the hottest air is. These automatic passive devices use no electricity and are useful if the homeowner does not want to be tied down to operating ports or if the openings are too high to reach conveniently. TACDs should be removed at the end of the heating season so that summer heat build-up will not be distributed into the house. The same port operators can be moved to outside vents during the summer and returned to common-wall ports in the fall. Homeowners reported very inconsistent performance in TACDs, so it is a good idea to get a recommendation or see one in action before making a purchase.

A thermostat can also cut the tie that binds a homeowner to a manually operated active heat-distribution system. A "cooling thermostat" will turn the fan on when the temperature in the sunwing rises above a predetermined point (unlike the heating thermostat attached to the forced-air furnace fan that clicks in when the temperature falls below a set point). There are many types of thermostats and automatic control systems available. A sunwing could have enough ports for manual passive convection as well as a thermostatically controlled fan system to take over when no one is home. There are fans that automatically increase speed as the air temperature rises, running more slowly and quietly at low temperatures. Or an axial fan could be used to blow warm air into the house on sunny days and reversed to blow warm air from the house to the sunwing at night and on cloudy days. The possibilities are limited only by the homeowner's ability to pay the HVAC piper.

Although these rules of thumb can get a homeowner started in designing a

THERMALLY ACTIVATED
CONTROL DEVICE

heat-distribution system, it is always wise to contact a solar professional before finalizing the system. As Ottawa designer Jonathan Cloud notes, "One of the primary considerations in attaching a sunwing is ensuring appropriate thermal cycling between the addition and the parent building."

BACKUP HEAT

Norm and Doreen Hutton heat their sunwing four different ways. On sunny days, the solar gain provides all the warmth plants need and some for the house too. In the evenings and on cool winter days, the wood stove in the family room next door takes off the chill, and during the night, electric baseboard heaters keep temperatures above freezing. When the Huttons are away on vacation, they connect two ducts from the forced-air propane furnace for reliable heat that needs no supervision.

Although most sunwings can collect enough heat to share with the main house on sunny days, in most parts of the country, they will not be able to maintain comfortable temperatures for either plants or people during extended

cloudy periods, cold snaps or the very short days of the winter solstice. In a solarium-sunwing, the need for backup will depend on the homeowner. If the space is to be used only when the sun shines, no backup may be needed at all, but a supplementary heat source will provide the flexibility of creating comfortable temperatures at any time without the cost of maintaining a stable minimum. A portable heater to warm the room occasionally may be all that is required.

A greenhouse-sunwing cannot tolerate temperatures below a certain level, regardless of outside conditions. In all but the coastal provinces and the southwestern niches of Ontario, a growing space will need a permanent, automatic backup heating system to maintain consistent reliable minimums to prevent freezing and to encourage growth. Because this represents the major operating cost of the sunwing, choosing the right heating system is important.

The simplest approach is to use **household heat** to maintain minimum lows in the addition. Reverse passive thermosiphoning can be encouraged by leaving doors and windows open at night or by omitting back-draught check dampers on the exhaust and return vents. Such systems are not particularly reliable, as one indoor gardener discovered the hard way. She usually left the door to the sunwing open at night to keep the space above freezing in midwinter, but on Christmas Eve, with all the festivities, she forgot. The next morning, she found her carefully nurtured plants frozen stiff.

Many homeowners connect ducts from the home heating system directly into the sunwing. This approach is not recommended as a permanent solution because the thermal cycling of a

With no insulated walls, this fully glazed sunwing is expensive to heat. As a result, it gets only seasonal use and is shut off from the house in winter. Terry and Sherry Marcotte, North Bay, Ontario

CUTTING GREENHOUSE
HEATING COSTS
Instead of maintaining
growing-level temperatures
in the whole room, a
dropsheet draped over the
growing beds will reduce the
volume of air to be heated.

A 'GREENHOUSE' WITHIN A GREENHOUSE CAN REDUCE THE VOLUME OF AIR THAT NEEDS TO BE KEPT WARM OVERNIGHT

DROP-SHEET

FORCED-AIR HEATER

sunwing is not in sync with that of the rest of the house. A sunwing heats up quickly during sunny periods, when the main house may still need heat, and cools down quickly, calling for heat long before the house needs the furnace on. The minimum low required for human comfort is warmer than necessary for plants, which generally prefer a cooler environment than people. The only way the house and sunwing can reliably and effectively share a heating system is by

means of a common-wall fan that is thermostatically sensitive to sunwing temperatures, drawing warm air from the house when the addition cools to a preset point.

The most practical solution is a sunwing heat source that is independent of the central heating system of the house. This **zone heater** can be a permanently installed wood, natural gas or propane combustion heater; an electric baseboard heater; or a portable

unit, electric or kerosene. The choice will depend on local fuel prices. The size will be determined by the local design temperature, since the backup must be able to keep the space at the desired temperature during an extended sun "drought." Whatever the heat source, it should be equipped with a fan, for during cold weather, strong convection currents are set up along the glazing, and only a forced-air system can ensure that the whole sunwing is evenly warmed.

Wood stoves are particularly suitable as backup heaters for solarium-sunwings. "Wood has the advantage of providing radiant heat, which offsets the glazing's negative radiant loss," says Michael Lambert of Solterra Developments Ltd., in Ottawa, "and it is usually cheaper." Although it is often too hot and too inconsistent for plants, wood heat can be used in conjunction with thermal mass to dull temperature extremes in a greenhouse-sunwing. John Przewoznik set two black, 50-gallon drums in a fieldstone wall separating the northeast corner of his greenhouse-sunwing from the adjoining room. A wood stove on the other side heats the house and passively charges the stone/water storage at the same time.

Like wood, combustion-type unit heaters, such as propane or natural gas, require a flue or external vent and an adequate supply of combustion air. If an outside air supply is not provided, combustion will cause too much negative pressure in the sunwing, sucking cold air in around the glazing and warm air from the main house.

On heating systems other than wood, thermostatic controls can automatically maintain sunwing temperatures above a certain set point. Serious indoor gardeners may want to invest in a temperature alarm that can be wired into the sunwing to give warning when the mercury is dangerously high or low, or when the backup heating system fails. In sunwings that do not require a minimum level of warmth, there should be a master on/off switch to isolate the sunwing during the coldest weather, if necessary.

The controls for backup heat must be carefully positioned so they accurately reflect the microclimate in the room as a whole. At the edges of the heating season, a daytime breeze might be welcome, but it could trip the supplementary heat on if the thermostat is near an open door or window. Conversely, controls installed in direct sunlight may register a high temperature even though the space is cool. On the north wall of her greenhouse, Mary Coyle has a thermostat that turns on a fan to extract warm air from the house when temperatures in the sunwing fall to 50 degrees F. In early morning, the sun strikes the thermostat and the fan shuts off, even though the greenhouse is still not up to temperature.

In a solarium-greenhouse sunwing, the minimum low maintained for plants will be too cool for human comfort. Yet it seems a waste of energy to crank up the thermostat to heat the entire space just to spend an hour reading the newspaper before supper. To solve this problem, the Donnellys have a radiant "people warmer" mounted on the wall behind a favourite chair. When that is turned on, the area around the chair basks in warmth: comfort at relatively low energy cost.

Backup heating costs will be most significant in a greenhouse-sunwing. In areas with long, cold, cloudy winters, the homeowner may want to experiment with energy-conserving approaches such as root-zone heating, an infrared system that warms the plants directly without heating the air, and localized heating

"It has become quite a chore to put them in every night," says this homeowner of his 22 panels of rigid insulation. "But it does the trick." With the panels in place, the addition gains more heat in the daytime than it loses at night. Max and Edith van den Berg, Winnipeg, Manitoba

Robert Tinker

such as tents or covers that create a greenhouse within a greenhouse and drastically reduce the volume of air that needs to be warmed.

In the milder parts of the country, however, even greenhouse heating costs will not be significant if the sunwing is well designed. Orval and Julie Meisner of Queen's County, Nova Scotia, designed their 240-square-foot greenhouse-sunwing with a solid roof, 55-degree sloped glazing, insulated foundation and 3-inch thermal mass floor. "We have plants growing all year long," report the Meisners. "We thought we might have to close it down for December and January. Instead, we are able to pick tomatoes and peppers." They spend $30 to $40 a year on the intermittent operation of a small quartz heater on very cold nights and on a small power outlay to run the electric reversible fan in the common wall, but the minimal operating costs are offset by

the excess heat gain pumped into the house. "After a few winters, it appears we are saving one to two face cords of firewood, mostly on late-autumn and early-spring days when no other heat is required. It has performed even better than our expectations."

VENTILATION

A sunwing's ventilation system is the summer counterpart to its winter heat-distribution system. Instead of transferring excess solar gain into the house, the ventilation system exhausts it to the outside, cooling the sunwing. The ventilation system can also be used to cool the main house, although it will likely be more comfortable anyway, because the addition acts as a sunblock for the south face of the house. Since very few Canadians rely on air-conditioning, the summer heat exhaust system does not offer much potential for energy savings, but if the homeowner

wants to use the space in warm weather, especially for growing plants, a good ventilation system is vital.

Because of the high azimuth angles around the time of the summer solstice, venting may be most needed in late spring and early fall, when the sun's rays penetrate the glazing more directly and ambient air temperatures are still high. In the regions that enjoy a continental climate, ventilation requirements will be easily met, since even the hottest days are usually followed by cool, fresh evenings that mitigate against uncomfortable heat build-up.

Ventilation systems offer the same four options as heat-distribution systems: they can be active or passive, manually or automatically controlled. Many of the same airflow principles also apply. Exhaust vents allow warm air to escape to the outdoors, so they must be positioned close to the peak of the sunwing where the hottest air

EXHAUST VENTS
Wind patterns on the site will determine the position of vents to exhaust unwanted solar heat from the sunwing. Because they are holes in the building shell, vents must be planned carefully to minimize winter heat loss.

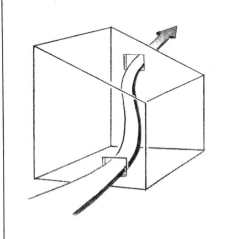

If prevailing winds are from west, vents are positioned like this.

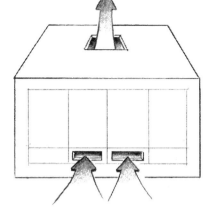

If summer winds are from south, vents must be in south glazing.

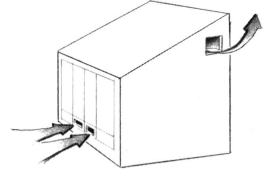

Exhaust vents high in side wall are preferable to roof vents.

accumulates. Whether the exhaust vents are installed high in the roof or in the east or west wall depends on the direction of prevailing summer winds — warm air must be exhausted on the leeward side of the building. If summer breezes waft from the south, exhaust vents can be installed in the roof. Although common in prefabricated greenhouse and sunspace kits, roof vents should be avoided wherever possible, for they are hard to seal properly and become a source of heat loss in winter. Operable skylights have the same limitations, only at greater expense.

Cut the intake vent, which draws cooler air from outside into the sunwing, in the wall opposite the exhaust vent, on the windward side of the sunwing. Position it low in one of the end walls or the kneewall (in shade, if possible), and separate it vertically from the exhaust vent by at least 6 feet.

According to the *Seymour-Dorgan Report*, for passive ventilation, the total area of vents in the walls and roof should be no less than 5 percent of the glazing area in order to maintain sunwing temperatures within 12 to 18 F degrees of outside air temperature. The percentage increases gradually in proportion to the amount of cooling desired: for a sunwing temperature within 5 to 9 degrees of the outside temperature, vents must be 22 to 28 percent of the glazing area. Solar Nova Scotia recommends that if the doors in the end walls are the only vents, they should be at least 30 to 40 percent of the floor area of the sunwing.

Exhaust vents should be larger than intake vents. Although 2:1 is a workable ratio, some designers recommend that upper vents be three times as large as lower vents. The resulting negative air pressure will speed up the rate of air exchange in the sunwing. Depending on

the size and configuration of the space, more than one exhaust or intake vent may be necessary. To reduce air infiltration, a few large vents are preferable to a series of smaller ones with a greater total perimeter.

"If there is low-slope glazing with built-in overheating, the entire end walls may have to be given over to vents,"

suggests Roscoe. Indeed, in the Hodson sunwing (45-degree slope), he designed an exhaust vent opposite the intake for cross-ventilation across the growing bed, a high exhaust vent to get rid of peak air, and a door — all on the west wall. Even with hot tub and concrete storage, the three vents have to be wide open to keep temperatures down in spring and fall.

Patio doors at each end of this sunwing provide enough cross-ventilation in summer to keep the room comfortable on even the hottest days. Rory and Christina Dowler, Russell, Ontario

Jim Merrithew

61

SUMMER COOLING
Although it is a solar
structure, a sunwing can
also "air-condition" a house
during warm weather if
airflow paths are planned
in advance.

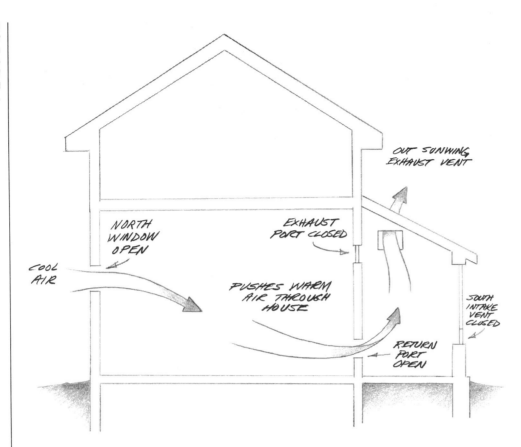

NORTH
WINDOW
OPEN

COOL
AIR

OUT SUNWING
EXHAUST VENT

EXHAUST
PORT CLOSED

PUSHES WARM
AIR THROUGH
HOUSE

SOUTH
INTAKE
VENT
CLOSED

RETURN
PORT
OPEN

country that are plagued with blackflies and mosquitoes during warm weather, a properly ventilated sunwing can be as useful in summer as it is in winter.

The intake and exhaust vents encourage convective currents that cool the sunwing, but the exhaust vent can also be used to cool the house itself, providing there is relatively free airflow between the sunwing and the opposite side of the house. Cool, shaded air that blows in a north window will flow through the house, pushing warm air ahead of it through the open return ports in the common wall and out the exhaust vents in the sunwing. (The exhaust port and the intake vent are closed.) Once the sunwing is built, it may take some experimentation to discover the right combination of open windows, ports and vents, but it is worthwhile to consider existing house windows and airflow paths during the design stage.

Cooling can be accelerated with fans, though it should not be necessary to resort to active ventilation in a solarium-sunwing with properly sized vents and solar controls. However, an active system may be necessary in a greenhouse-sunwing, since the sloped glazing can create serious heat build-up. The fan system should be based on the hottest weather during which the greenhouse will be operated (up to 100 air changes per hour may be needed). This active ventilation will also help reduce high summer humidity. Position fans so they do not create draughts for the plants: after her sunwing was built, Mary Coyle had to fit one of the fans with a channel to direct the airflow away from growing beds.

The advantages, disadvantages and sizing of fans for exhausting summer heat are basically the same as for winter heat distribution. The only difference is that since the summer fan is exhausting

Vents may simply be openings framed in the outside walls and fitted with insect screens and dampers. If the sunwing will be used during the heating season, however, special attention must be paid to their design and construction. Unless the vents can be closed off permanently and sealed tightly at the end of the cooling season, they will be a major source of heat loss as winter approaches. Vents in solid walls should be fitted with weatherstripped, movable insulation that raises the R value to the same level as the wall they perforate — outside shutters are best. Although off-the-shelf operable windows are more expensive than a hole in the wall, they

are preferred by many designers over specially constructed vents because they close tightly and can be easily shuttered from the outside. (They also come equipped with screens.) The windows should open fully and be installed so they "scoop" prevailing winds without admitting rain.

Sliding patio doors can also be used to improve passive ventilation and increase the sunwing's connection with the outdoors. The three 8-foot units installed in the south wall of the Donnellys' sunwing convert it into a screened porch in summer. "We've practically abandoned our outside picnic table," admits Gerald. In those parts of the

outdoors, greater consideration must be given to heat-loss potential. Fans to the outside should be inset enough that a rigid insulation panel can fit tightly over them in winter. If possible, try passive ventilation first: the expense and noise of fans can always be added later. The homeowner can cover both options by building in passive ventilation but framing and wiring the appropriate wall sections to take a fan if one becomes necessary. This adds little to the cost and is easier to do during construction than after the fact.

Controls for summer venting can be either manual or automatic. For manual control, the indoor-outdoor thermometer that provided winter readings between sunwing and house can be moved (or another installed) so the temperature differentials between the sunwing and outdoors can be monitored. The readings will indicate when to open vents before indoor conditions become stifling. Manual controls range from relatively sophisticated rack-and-pinion systems to a pole on a slotted board. Designing manual controls for high exhaust vents can sometimes be a challenge. Place them where they are accessible from a ladder, balcony or operable side window, if possible. The Donnelly sunwing wraps around an upstairs balcony onto which two upper vents conveniently open.

Although solarium-sunwings can be ventilated manually, automatic vent openers are a good investment for greenhouse-sunwings, which are prone to overheating. Unless an indoor gardener is certain to be home at all times, only automatic venting can protect his plants. The TACDs that operated the common-wall ports in winter can serve double duty on the outside vents in summer. Be sure to shield the piston on the intake vents

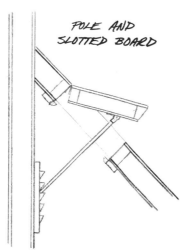

POLE AND SLOTTED BOARD

from breezes that will make it close prematurely. Most TACDs do not have the force to pull the vent tightly closed, so in the shoulder seasons, when it may be necessary to vent during the day and heat at night, the homeowner will have to shut the TACDs manually to prevent heat loss. Vents may also have to be weighted against high winds, especially if they open sideways.

For a more sophisticated system, vents can be operated by a thermostatically controlled motor that opens and closes vents by means of a gear-and-chain drive. The first automatic thermostat, called the "automaton gardener," was introduced in England in 1816, but today, there are many varieties available. An adjustable "open, close and limit" thermostat will open and close vents when the temperature rises above or falls below certain set points. Modulating thermostats open and close vents gradually as the temperature inside the sunwing varies. Of course, TACDs and thermostatic controls must be turned off or disconnected during the heating season.

Although many plants suffer from the

heat, people can generally withstand high temperatures, especially after enduring the lows of winter. Elizabeth Cran reports that her P.E.I. greenhouse-sunwing has frequently reached 150 degrees F, and though she is trying to improve conditions for the sake of the plants, she does not mind the heat herself. "However hot it is, it doesn't feel unbearable in the greenhouse."

SOLAR CONTROL

Controlling summer heat collection is an important corollary to ventilation: if the sun's rays can be blocked *before* they enter the room, there will be less heat for the vents to exhaust. Designs with south-facing sloped glazing will need solar controls most, but west-facing windows will also be prime candidates for summer sunblocks. Although few of these solar-control systems are part of sunwing design, per se, they are integral to the overall planning and cost of a passive solar addition.

Solar controls fall into roughly three categories, depending on whether they work on a permanent, seasonal or daily basis. **Permanent shades**, such as solar-control films, limit the amount of incoming solar radiation year-round. These are definitely not recommended for south-facing glazing. Although it is not always clear from the product advertisements, any sunblock that is permanently applied to the glazing will cut valuable winter radiation just as much as unwelcome summer sunshine. Permanent solar-control films might be considered for west windows, which supply little solar gain in winter yet are a source of summer overheating.

Seasonal controls make more sense for south glazing, blocking the sun only when it is hottest. The cheapest option is shading compound that is brushed or painted on the glazing in spring and

MANUAL VENT CONTROL
The simplest and cheapest way to exhaust summer heat is a site-built vent propped open with a pole and slotted board. However, this method is only as effective as the person operating it.

63

OVERHANGS
Though it effectively
prevents summer
overheating, a structural
sunblock is not as
appropriate for northern
latitudes because it blocks
the sun equally in April and
August, even though solar
heat is welcome during a
Canadian spring.

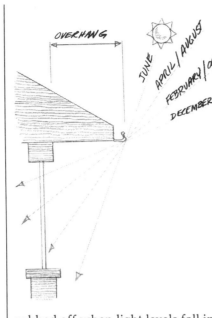

ADJUSTABLE LOUVRES
An improvement over solid
overhangs, though more
labour-intensive, louvres
can be tilted to admit or
block the sun, depending
on the sunwing's heating
needs.

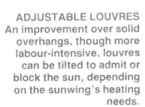

rubbed off when light levels fall in the autumn. There are many variations of the concept available from commercial greenhouse suppliers. In general, the thickness of the coat determines the amount of radiation that is blocked. Some wear off gradually through the summer; others claim to stay on even in heavy rain but will rub off with a dry duster. Another form of paint-on shading is opaque and white when dry but becomes translucent when wet, allowing more light to enter on rainy days. A traditional homemade shading compound can be made by mixing white lead, kerosene and linseed oil. The first layer is applied in early spring, with more layers added as the sun's intensity increases. If such paint-on shading compounds are used, be sure they are compatible with both the glazing material and the mullions between.

Structural sunblocks such as overhangs are less messy and time-consuming. These seasonal shades are permanent, designed to coordinate with

the sun's path, blocking radiation when the sun is high in the sky but not interfering when the sun is low. Because of the low angle of the rising and setting sun, overhangs are ineffective on east and west glazing, and even on south glazing, they are more applicable in the southern United States than in Canada. For instance, in New Orleans, 1 foot of overhang puts 5 feet of a vertical south wall in shade, whereas in Saint John, New Brunswick, it creates only 2 feet of shade.

Because it provides maximum shade at the summer solstice, around June 21, it is impossible to design an overhang to shade the south glazing in July and August without eliminating useful spring heat gain. The problem is that in this country, the sun and the human need for heat are drastically out of phase with each other. As Alberta designer Michael Kerfoot says, "Around here, August is swimming-hole weather, whereas in April, you've just finished ice fishing, yet the sun's position in the latter part of April is the same as in late August!"

Although sunpaths do not always coincide with the sunwing's need for heat, vegetation often does. Because June, July and August are the primary growing months, plants create seasonal sunscreens at exactly the right time of year. Plants that die back each fall are more useful than permanent perennials, such as trees, that block some winter sun. Fast-growing varieties such as sunflowers or cosmos can stretch 6 to 8 feet in a couple of months. Vines of morning glory or scarlet runner beans can be trained to grow up the glazing or along overhead trellises, creating shades that are not only effective and colourful but, if the plants are perennial, automatic. One of the best shade plants is hops. Gary and Sandy Hodson can use their sunwing all summer, despite its

extremely low slope, because from mid-June, the glass is blanketed with the brewer's vine. In spring, they nail pea netting over the glass and simply remove the plant-entwined screen in late fall. The hops conveniently confine themselves to the glass because the surrounding asphalt shingles are uncomfortably hot.

Although seasonal controls function well in midsummer, only **daily solar-control devices** — blinds, screens, shades and awnings — can effectively modulate the amount of heat gain in spring and fall. Interior blinds are the least effective, for they intercept heat energy after it has already penetrated the glazing. If the blind is dark-coloured and there is a space between it and the glazing, a very efficient solar collector is created: cool air is sucked in at the bottom of the blind, rising and gathering heat until it blows warm air out the top. Convective currents will be minimized if

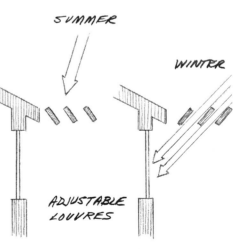

the blind lies close against the glazing and is sealed to the window frame, top and bottom, but then the inside pane of glass absorbs the incoming heat energy. If it heats up proportionally more than

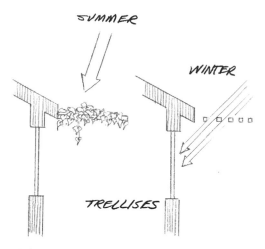

SUMMER

WINTER

TRELLISES

the outside pane, the stress of differential expansion can break the seal between double insulating glass units. For this reason, interior movable insulation should never double as a solar control, unless the window side of the blind is highly reflective to avoid such heat build-ups.

Although not as readily available, shading devices that are installed between two panes of glazing are an improvement over inside blinds. Any heat absorbed by the shade is transferred to the airspace and vented to the outside so that convective currents actually work to cool the glazing rather than to increase the heat gain.

Exterior shades are the best choice for daily controls, since they stop radiation before it reaches the glazing. There is an infinite variety of daily shading systems — blinds, netting, plastic sheeting, strips of muslin. The cheapest external shade is a plastic cloth made of weather-resisting and nonrotting fibreglass, polypropylene or other synthetic material. It comes in a wide range of colours with various light-blocking capabilities. Shade cloth that blocks 70 to 80 percent of the incoming insolation works well for solarium-

sunwings, but in a greenhouse-sunwing, sun control must be balanced against the plants' need for light: no more than 25 percent should be blocked. It can be draped, stapled, laced or clamped onto the outside of the building or fixed onto rollers at the ridge of the greenhouse-sunwing and controlled with pull cords.

There should be an airspace between the shade and the glazing so that the outside lite (or pane) is cooled by convection currents. Otherwise, differential expansion can cause seal failure in double insulating glass units. Without an airspace, the shade cloth itself becomes extremely hot and could permanently bond to some types of synthetic glazing. The drawback to outside shades is that they are exposed to outside conditions. They must be installed securely, especially in windy regions, and may need replacement after a few seasons of the rain-sun cycle typical of summer.

A more durable shade that filters sunlight, rather than blocking it evenly, is an external roller blind of redwood slats, bamboo or aluminum. Although the natural material and rhythmic patterns of wooden slats are attractive, they may look a little bedraggled after being exposed to the weather for a season or two.

Whatever daily shade control is chosen, the operating mechanism should be both sturdy and convenient. If it is not easy to pull up and down, the shade will be little more than window dressing. An inexpensive sun control that the homeowner feels comfortable using will be better in the long run than a sophisticated system with difficult controls.

HEAT-LOSS CONTROL

The glazing that creates unwelcome heat gain in summer becomes a major

source of heat loss in winter. This "hole" in an otherwise well-insulated shell lets out more than its share of precious warmth. If there is an equal area of insulated wall and glazing in the sunwing, with respective R values of 20 and 2, the average insulating value of the building skin is not an acceptable R10, as many people think — it is closer to R4! Unfortunately, an effective, economical and convenient way of controlling heat loss through sunwing glazing has thus far eluded cold-climate designers.

Heat energy is lost through windows in three ways: it is **conducted** through the glazing material; moving air carries it from warm parts of the room toward the

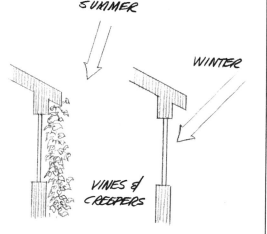

SUMMER

WINTER

VINES & CREEPERS

cool glazing surface by **convection**; and the warm objects in the room constantly **radiate** heat energy toward cooler objects, like glazing, because it is the nature of heat to seek an equilibrium. Insulation slows down conductive losses by trapping heat energy in dead airspaces; convective losses are cut by stifling air currents; and radiant losses are reduced by blocking with a low-emissivity shield. Those fundamental principles have spawned countless

TRELLIS
Plants provide sunblocks in midsummer and early autumn when they are needed most, though the supporting framework of a trellis may create some unwanted winter shade.

VINES
The best seasonal sunblock is vegetation that grows on disposable netting and dies back to the ground each fall.

variations in heat-loss control systems for glazing: they can be put inside, outside or between the panes; they can fold, roll, swing and slide; and they can be opened and closed by thermostats, cranks, pulleys or human hands.

Although the nature of winter in most parts of Canada mitigates against them, outside shutters are the most effective and least problematic window insulation from a thermal standpoint. They prevent severe cold from ever reaching the glazing, thus protecting it from the extreme temperature changes that test seals to the limit. The room-side pane remains warm, eliminating the problem of convection currents and condensation. Shutters also protect the most vulnerable (and most expensive) part of the sunwing from mechanical damage. Although plodding through 3 feet of snow every day to open and close shutters is not convenient, outside insulation is the best solution for windows and vents that are sealed for the whole heating season. Chris Jalkotzy used this seasonal heat-loss control for the basement sunwing he designed for his parents' home in Fort Smith, Northwest Territories. Through the winter months, the single glazing is snugly blanketed with rigid foam insulation shutters.

Movable insulation can also be installed between the lites of a double-glazing system. Zomeworks developed Beadwall, in which zillions of tiny beads of rigid insulation are blown into the cavity between glazings, turning the

Roger Vernon

window into a wall at the flick of a switch. There are also commercially available systems with rigid insulation sliding shutters or roll blinds between the two panes of glazing. This approach makes sense structurally, since the insulation is sandwiched between an inside glazing that stays warm and an outside glazing that protects it from the weather. An exterior access panel above the glazing should be provided for maintenance and repairs. Although commercial interglazing systems are expensive, Michael Kerfoot feels this is a realistic option for the owner-builder. "Putting insulation between the glazings is easy to do at the design stage, and there is little extra cost to building in the pockets and operating systems."

The bulk of commercial and do-it-yourself systems put window insulation where it is most convenient for daily operation – *inside the glazing*. Yet this is where it is least effective. In fact, curtains and blinds with poor perimeter seals can actually increase heat loss by creating strong convection currents that suck warm air toward the glazing. On midwinter nights, if the surface temperature of the glazing dips below the dew point, infiltrating room air will deposit condensation, creating moisture problems for the sills when it melts next morning. If the insulation works, the inside pane is put through a gruelling daily cycle of expansion and contraction as the glazing warms and cools. Although some heat is saved, the window itself might be lost: if the glazing is deeply chilled overnight, then warmed very quickly by intense sun while the insulation is still in place, the thermal stress of differential expansion can cause seal failure in insulated glass units. If the blind is opened suddenly, exposing the cold glass to warm air, it can crack from thermal shock.

Though not as effective at reducing heat loss, insulating window blinds are more convenient to operate than exterior shutters. Eric Darwin and Frances Dubois, Ottawa, Ontario

Sealing curtains and blinds to the window frame is difficult in most situations, and particularly challenging if the glazed area is large or sloped. Most variations are suspended from guide wires or run in tracks attached to or inside the window frame. It is difficult to design a system that moves up and down smoothly yet seals tightly along all edges.

Rigid insulation is more effective because it can be pressed tightly enough against the glazing to stop air movement and can be made to a higher R value, though some rigid insulation is prohibited in exposed areas because of potential fire hazards. Sealing the edges against infiltration is still a challenge, as is designing a system for large expanses of glass, since joints are susceptible to heat loss. Rigid insulation panels can also be heavy if very large and are more awkward than blinds to operate as well as to store. Although they need space to swing open, hinged inside shutters are relatively successful for smaller

This insulated interior shutter represents a compromise between the homeowners, who wanted a view from the east side of their sunwing, and their architect, who disapproves of side wall glazing. Gerald and Joan Donnelly, Albion Hills, Ontario

Gail Harvey

windows. John Hix specified plywood-covered rigid insulation for the east and west windows of the Donnelly house. Well weatherstripped to stop infiltration, the windows remain shuttered for most of the winter. They are hinged on the north side so they can be opened in warm weather and hooked unobtrusively against the back wall.

Because every sunwing is unique, window insulation must be custom-made, which makes it very costly. Homemade designs, including blinds, curtains, Roman shades and pop-in or hinged shutters, are less expensive and not difficult to make, but they are tricky to install with effective edge seals. Both commercial products and do-it-yourself plans make R-value claims for their products that will probably not be realized in actual operation. The combined heat resistance of the materials is usually compromised by heat lost around the perimeter.

Potential energy savings, however large or small, will depend on whether or not the movable insulation is actually used. "Window insulation is a genuine bother to put up and take down on a daily basis," says Martin Foss, who abandoned the idea of fabric window insulation after one winter's trial. Manual operation demands considerable commitment from the homeowner, and automated systems, though reliable, are costly. Because they will be subjected to daily use for almost half the year, operating mechanisms should be both easy to use and very sturdy.

"The movable insulation systems at present commercially available," concludes Michael Glover, "are not cost-effective and do not perform and operate satisfactorily. From a heat-loss standpoint, energy dollars are better spent on additional layers of glazing or better-quality windows." The trend among Canadian designers also seems to be away from movable insulation. As John Hix says, "To try to cover all that glazing with movable insulation is expensive, and you run into the problem of creating a condensation sandwich." He prefers to close the sunwing off from the house and let it cool down. The windows do not have to be insulated because the heat lost through the glazing is not replaced with purchased energy.

Even if the homeowner pays for replacement heat, movable insulation is not cost-effective, according to Bob Argue: "Once you have a well-built, well-insulated house with windows well oriented, the percentage of heat-loss reduction is not that great. The fuel dollars saved probably don't warrant the high cost of movable insulation. All you need is something to stop the radiant 'coolth' to improve the comfort level."

When it is colder outside than inside, the bare windows attract radiant heat from the warm objects in the room, making anyone sitting nearby feel chilly. Any window covering that cuts this radiant loss — a curtain with a reflective backing or a sheet of Mylar — will make the room feel more comfortable, even if overall heat loss is not significantly reduced. This explains why many homeowners are satisfied with their investment in inside movable insulation. Even though interior systems do a poor job as insulation, the cost is justified if they serve other purposes — blocking summer sun or neighbours' prying eyes, improving comfort or adding to the decor. "Ideally, you should have four separate window systems for heat rejection, heat retention, privacy and aesthetics," says Michael Kerfoot. "But that is hardly practical. The best solution is a single window system that does a reasonably good job of meeting as many of those needs as possible."

Homeowners who decide to go ahead with window insulation should calculate that as part of the budget and accommodate rolled-up curtains, swinging shutters or storage for stacks of insulating panels in the design. Plan glazing details according to the dimensions of the insulating material. If inside blinds are designed to function as shade controls, their outside surface should be solar-reflective, and they should not be sealed tightly when the sun is shining.

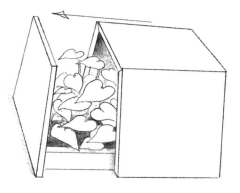

Heat-loss control is most important in greenhouse-sunwings, where the homeowner has to pay to keep the space at a consistent minimum. A good alternative to blanketing an expanse of glazing is a **growing bed cover** that controls heat loss directly around the plants, instead of from the room as a whole. Such thermal tents may be enough to keep plants warm during spring and fall without backup. In winter, this on-the-spot insulation could conceivably be coordinated with a local heat-distribution system that warms the growing beds but not the sunwing. Luc Muyldermans combines movable insulation over his plants with a solar-heated air delivery system to the soil. As a result, he can grow vegetables all winter with no backup, even in Quebec.

Sunwing Design Guidelines

VENTILATION

Exhaust vent high on leeward side; intake vent low on windward side

Avoid roof vents if possible

If roof exhaust vents necessary, position intake vents in kneewall

Offset vents and separate vertically by 6 feet

Total vent area should be no less than 5 percent of the glazing area, with exhaust vents larger than intake vents (2:1)

Fit vents with insect screens and seasonal movable insulation

Use operable windows as vents where possible

Seal specially constructed vents very well

Passive/manual vents for solarium-sunwings

Active/automatic vents for greenhouses used during warm months

Consider house airflow when designing ventilation system

GLAZING

Oriented within 30 degrees of solar south

Unshaded during winter between 10 a.m. and 2 p.m.

Maximum south glazing for heat collection or plants

Sloped glazing for greenhouse-sunwings

Vertical glazing for solarium-sunwings

Skylights for improved solarium/house daylighting

Solid end walls for solarium-sunwings

East glazing for greenhouse-sunwings

Common-wall glazing for solarium-sunwings

HEAT DISTRIBUTION

Passive distribution through ports in common wall

Ports offset and separated vertically by 6 feet

Exhaust ports one-third larger than return ports

Return ports from cooler part of house

Total port area at least 2.5 percent of glazing area

Back-draught check dampers to prevent reverse thermosiphoning

Supplement passive convection with fans if large glazed area, sloped glazing or little mass

Fans in greenhouse-sunwings to combat stratification

Manual controls for solarium-sunwings

Automatic controls for greenhouse-sunwings

THERMAL STORAGE MASS

Use thermal mass to temper extremes, not to provide cloudy-day and night-time heat

Incorporate thermal mass if the system is simple and inexpensive

For buffered, unheated sunwings, integrate thermal storage in structural and finish materials

For passive absorption, expose storage mass to direct sunlight, and use a rough, dark-textured finish

For heated sunwings in mild climates, use integrated hybrid storage (active delivery, radiant recovery)

Distribute storage mass evenly

By far the most convenient method of reducing overnight heat loss through the windows, tracked curtains help keep this sunwing at livable temperatures all winter long with no backup heat. Greg and Linda Powell, Priddis, Alberta

Roger Vernon

70

throughout the interior for effective radiant recovery

Insulate storage mass underneath and around perimeter

BACKUP HEAT

Optional for solarium-sunwings; mandatory for greenhouse-sunwings

Direct connection to central heating not recommended

Independent "zone" heater with separate thermostat recommended

Forced-air heater to combat convection currents

Wood stove ideal in solarium for intermittent use, especially combined with storage

Automatic controls for greenhouse

Position controls where unaffected by direct sun or draughts

SOLAR CONTROL

Avoid permanent sunblocks on south glazing

Solar-control film appropriate for west glazing

Calculate overhangs carefully to avoid spring shading

Use vegetation shade where possible: trellises, bushes, tall plants, vines

Install daily sun-control devices on exterior

Provide airspace between exterior shade cloth and glazing

HEAT-LOSS CONTROL

Evaluate cost and benefits of window insulation carefully

Provide outside seasonal shutters where applicable

Be sure systems are durable, tightly sealed around the perimeter, easy to operate

Provide radiant-reflective curtain or simple drapes as alternative to movable insulation for large south windows

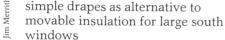

Integrating the sunwing with the house can be accomplished by choosing decorative details that are aesthetically suited to the style of the house. Eric Darwin and Frances Dubois, Ottawa, Ontario

5

Taking Flight

Plans, prefabs and payments

'Always look before you leap. The conclusion you jump to could be your own.'

— James Thurber

PLANS

With the help of her sister and as many books as she could find, Elizabeth Cran drew up her own plans for a sunwing addition to her 19th-century Prince Edward Island farmhouse. She knew what she wanted, and the builder she hired seemed to understand the elementary drawings she had sketched. However, when he built the addition, he fastened the floor to the top of the foundation wall instead of 2 feet below, as Cran had specified, eliminating the kneewall that was to give the plants headroom. Although she has adapted to the error by only growing plants that sprawl into the narrow triangle, a better set of plans would have given Cran *exactly* the sunwing she had had in mind.

Between design and construction is the all-important stage of drawing the plans, transforming those idealistic doodlings into blueprints for an actual sunwing. Many people will want to see the design on paper — the municipal officials who oversee local bylaws, the inspector who enforces building standards, the bank manager and the contractors or tradespeople bidding on the job. The plans for City Hall have only to be accurate enough to show what will be done and to provide an estimate of the cost of doing it. For instance, city officials will want to know that there is to be a small casement window in the east wall, but they may not need to know the final dimensions. The detailed drawings, on the other hand, are the builder's construction guide and must be absolutely precise: if the air/vapour barrier is omitted on the drawings, it may be omitted from the wall cavity as well. Even if the homeowner designer/builder is the only one to see them, a set of plans should be prepared: it forces one to think through each sunwing detail in advance and to clear up problems before construction begins. The more detailed and accurate the drawings, the better the homeowner's chances of ending up with a sunwing that matches expectations.

A set of plans includes a site plan, a floor plan, interior and exterior elevations of all sides of the new space, a building cross-section and enlarged construction details, as well as a specification sheet that lists the materials, practices and responsibilities of the various people involved in the sunwing construction.

The **site plan** is an overhead view of the whole property, showing boundary lines and any zoning constraints that apply, such as easements and setbacks. The house, trees and other obstructions should be drawn to scale and positioned correctly on the lot. The current outside dimensions of the house must be indicated, as well as the new dimensions after the addition is constructed. The site plan shows at a glance exactly where and how far the sunwing will extend.

A **floor plan** is an overhead view that zeroes in on the sunwing and as much of the existing house as is affected by the new construction, usually just those rooms adjacent to the common wall. An original house plan may be available from the first owners of the house or from municipal offices. If not, one must be drawn using a scale of ¼ inch = 1 foot. Measure the inside and outside of the house wall where the addition will be attached, using a long tape to avoid "carpenter's creep" — the inaccuracy that comes from adding a series of short measurements. The scaled floor plan must show the horizontal dimensions (length and width) of rooms, halls, stairwells and door and window openings. Pencil in the function of each room and the location of all plumbing, vents and wiring. Indicate the size, species and spacing of floor joists in the addition, as well as any structural

Attsun Sun Space Systems

information relevant to special support systems, such as headers for new openings in the common wall or special beams to support sloped glazing.

Elevations show the vertical dimensions of the new construction and the edge of the house to which it is attached. There should be exterior elevations of the east, west and south views, as well as interior elevations of all four walls, including the common wall. Elevations indicate window and door positions and dimensions, inside and outside heights, the size of overhangs and degree of slopes. They also give material descriptions of interior and exterior finishes, the positions of vents and thermostats, and any construction details outside normal practices, such as unconventional framing or lateral bracing.

The **building section** is a side view of the addition showing all the structural elements of a building normally hidden inside the walls and underground. It shows the common wall and its foundation, the roof and floor structure and the south wall of the sunwing down to the footing. The cutting plane does not have to be a straight north-south line; it can jog if the draftsperson wants to indicate a particular detail. This cutting plane is marked on the floor plan with a dotted line.

The building section is not drawn to a very large scale and shows only the general arrangement of structural members and interior finishes. Like the floor plan and elevations, the building section will likely have certain areas circled and lettered or numbered, indicating that these **details** have been enlarged elsewhere to clarify a construction sequence, specify an unusual material or detail an unconventional construction practice. Enlarged details are most commonly used where there might be confusion about how to terminate materials such as air/vapour barriers or flashings or where there is a transition between

Garden Way Manufacturing Company

finishes. These details are important because they translate the designer's vision into the language of the builder.

Preparing plans is not beyond a novice's scope, any more than construction is. If homeowners are knowledgeable and meticulous, they can develop their own drawings, particularly if the sunwing is relatively small and the design is straightforward. To do a good job, homeowners must be able to think through each step of the construction and visualize the physical processes and problems that will be encountered on the job. Only then can they be solved on paper. If any of the above seems foreign, it may be better to hire an architect or draftsperson to do the work. The important point is not to skip this phase altogether and decide to play it by ear.

Even if someone else is hired to prepare the sunwing plans, reading them can be a chore in itself. Ask the person who drew them up to explain them in detail. If the plans were purchased from a mail-order supply, you are on your own. Although it is not within the scope of this book to give a course in blueprint reading, a few general directions may help. The working lines that delineate the actual shape of the addition will be the heaviest. Dotted lines are used to indicate something that is inside or behind another feature. For instance, an original door opening that is covered by the new stud wall would be shown with a dotted line. Dimensions are indicated with light, solid lines with arrow tips. Sometimes working lines are extended with light, solid extension lines so that dimensions can be shown more clearly. The scale of the drawing is generally indicated in the lower right-hand corner, and if none is shown, do not assume the plans were drawn to scale. The plans may have to be read several times before they make sense. Enlist the help of a

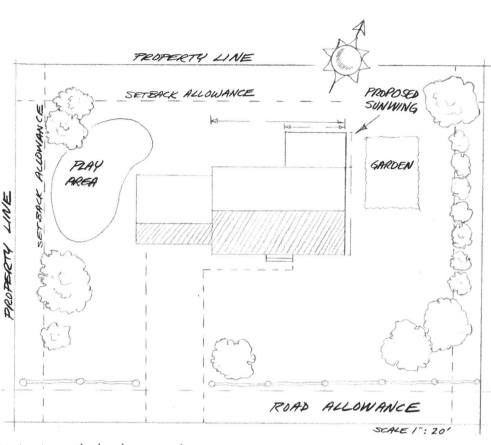

contractor or the local community college, if necessary, but do not proceed until everything is crystal-clear. Plans can be confusing and complex: going ahead without a firm grasp of every detail is an invitation to disaster.

Wayne Wilkinson had no trouble coming up with an efficient, attractive design for his Yukon sunwing. He is, after all, an energy consultant. Nevertheless, for final plans, he turned his preliminary sketches over to an architect. "I can't draw a straight line with or without a ruler, and I wanted specific structural input from someone with sensitivities to energy efficiency and solar design."

PROS

Architecture is peculiar among the arts in that it performs a practical as well as a purely aesthetic function. Homeowners may know what they want but not feel confident that their design is structurally and mechanically sound. Will the sunwing cut off light from the upstairs bedroom? Will the slope of the roof dump snow? Will an existing foundation support an addition? If the homeowners cannot fully answer such practical questions, they will need the services of a professional — an architect, designer, structural engineer, solar consultant, draftsperson or contractor. Consulting with a professional can more than pay for itself.

An architect or designer offers the fullest range of services, such as generating the design, preparing preliminary cost estimates, obtaining permits, hiring the contractor and supervising construction. He or she can be hired on a fee-for-service basis to do the design work only, shaping the homeowner's vague desires into a set of plans; the homeowner is then responsible for getting the sunwing built. The architect/designer can also be hired on an hourly consultation basis to walk through the house and make suggestions or to go over plans and point out potential structural, architectural or solar weaknesses in the proposal. A surprising amount of information can be gleaned by picking a professional's brain for an hour or two: it might be worth $100 just to hear an architect agree that the design for the $10,000 addition is on the right track.

The difference between an architect and a designer is one of accreditation. An architect is a registered member of the Royal Architectural Institute of Canada and thus has the right to put MRAIC after his or her name. Designer is a less precise term referring to anyone from an architecture-school graduate to an alumnus of the school of experience.

Legally, an architect must be consulted for projects over a certain size or dollar value, but otherwise, anyone can design a residential structure as long as it meets building-code standards. The services of a professional are probably not as expensive as assumed, but many additions are too small to interest an architect, and in any case, most architectural training is not strong on passive solar design.

John Hix splits the solar-design field neatly into three groups: the scientific types who run their designs through computers and do complicated calculations to find out how the solar design works and what the payback is; the house-and-garden group who are interested mostly in what the sunwing looks like; and the few who have actually lived in them. "Find one who knows how they work in real life," he advises. "Too few architects have experienced the designs they are producing. It is important that these principles are learned by living, in an empirical way. Otherwise, valuable ideas may be ruled out at the slide-rule stage."

Ideally, the chosen architect or designer will have experience in passive solar additions or will have incorporated those principles into new construction. Get recommendations from friends or the local chapter of the Solar Energy Society of Canada, Inc. (SESCI), or consult the yellow pages. Regional government offices that deal in energy conservation may know of knowledgeable people in the area. Mary Coyle tracked down her architect through the local community college; Eric Darwin was referred to his by a friend; Gerald Donnelly found his in a magazine; Hannes Jalkotzy had one in the family.

Arrange to have the most promising candidates visit the house to discuss the

THE FLOOR PLAN
This scaled drawing of the sunwing and adjacent rooms indicates all horizontal dimensions (length and width), the location of services, and structural information necessary for construction.

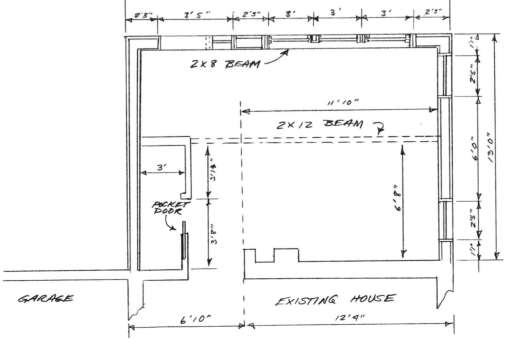

sunwing plans. Ask them for the names of previous clients who can be contacted, particularly those with additions similar to the proposed project. Find out what they charge and exactly what the cost includes. Enlarged construction details and specifications are not necessarily included in the plans.

"Draw up your own sketches and variations before consulting with an architect," advises Eric Darwin, who decided to hire a professional to make sure he wasn't missing something obvious in his design. "And get the plans drawn up long before you build so you can live them for a year and make modifications."

In many parts of the country, there are neither experienced solar designers nor energy-efficient contractors within a reasonable radius. The homeowner can, however, have plans vetted by mail by sending them to one of the companies listed in "Sources" (page 142) that will look over plans and make suggestions for a nominal fee. Then it is up to the homeowner to select a contractor who is open to new ideas and willing, perhaps, to do a little background research. As Bob Argue says, "The less experience there is available in a community in energy-efficient construction and passive solar design, the more experienced the homeowner has to be."

If the homeowner feels secure with the design or has it checked by a professional, a draftsperson can be hired to do the actual working drawings. And that is *all* he or she will do: don't expect a draftsperson to redesign the sunwing or pick up on ineffective or inconsistent solar design elements. This architectural drafting service will cost either an hourly rate or a fixed fee. Sometimes students from colleges or universities will do the drawings at a reduced rate, but under professional supervision.

Common-wall openings require careful planning to make them useful to both rooms. This window serves as a convenient pass-through from the sunwing and offers a pleasant view for kitchen cleanup. Eric Darwin and Frances Dubois, Ottawa, Ontario

Jim Merrithew

ELEVATIONS
Interior and exterior views of
each sunwing wall and the
common wall are drawn to
show the vertical
dimensions of the addition
as well as the position of
openings, services and
types of finishes.

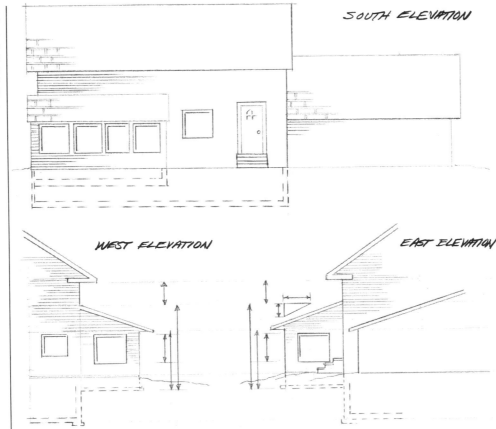

SOUTH ELEVATION

WEST ELEVATION

EAST ELEVATION

Paying a professional for design services can sometimes save as much as it costs, if only in avoided disasters. If the homeowner is the least unsure about any phase of the plans, it is well worth the expense to hire a professional.

PREFABS

Drawing up plans and dealing with designers does not appeal to everyone. To meet the demand for a worry-free, ready-made sunwing, several companies market prefabricated passive solar additions that arrive on the doorstep neatly packaged with a sheaf of installation instructions. Homeowners can erect the prefab themselves or hire a

contractor to put the kit together and attach it to the house.

The advertising brochures for these products show smiling families lounging amidst a veritable jungle of plants, and their claims are impressive: one model promises to "heat your home, feed your family, brighten your life and pay for itself." Prefabs certainly have the advantage of convenience, especially if the company that sells the kit takes care of its installation, but as with any purchase, prospective buyers should steel themselves to the appealing ads and carefully assess what they are getting for their money.

Most of the kits are modelled on the

traditional all-glass greenhouse, and though they now claim to be "for people, not just plants," the design was originally intended to provide maximum light for horticulture. Cold-weather heat loss and summer-heat ventilation were not important considerations in northern Europe and Britain, where fully glazed sunspaces were developed, but in most of Canada, such a fishbowl design will overheat unbearably in summer and will be extravagantly expensive to maintain at a livable temperature in winter. "Small glass sun rooms are appropriate for England, where the climate is less rigorous," says John Hix. "They are sold here as a bit of European heritage, but the designs are not applicable to our weather conditions."

Fully glazed prefabs may perform acceptably in southwest British Columbia or on the Maritime coast, but even in these mild climates, they should be double-glazed. Though some models are available with triple glazing, this option still offers only a fraction of the heat-loss resistance provided by an insulated roof or wall. Manufacturers' claims notwithstanding, there can be nothing energy-efficient about a fully glazed room. Double glazing and thermally broken metal frames slow heat loss somewhat, but if comfort-zone temperatures will be maintained year-round, expect a significant rise in home-heating costs. "With these spaces, it is best to adopt the 'sacrificial sunspace' mentality," says Michael Dorgan, an Ottawa solar-energy consultant. "Close them off from the house, and let them freeze — don't try to maintain a minimum low." For most of the country, that strategy severely limits the use of the space and turns the sunwing into an expensive, little-used showplace.

Although fully glazed plastic prefabs are no more energy-efficient than the all-glass variety and are also best used as unheated buffer spaces, they are much less expensive. Partly, this is because plastic glazing is cheaper than glass, thus cutting the price nearly in half, but another big saving is hidden underground. Glass, which is heavy and susceptible to damage from frost heaving, requires a full foundation, often costing as much as the kit, but a plastic prefab is lightweight and can be set directly on the patio or lawn, attached to a single row of blocks or 2 by 8s. Usually considered a temporary addition, it probably will not need a building permit and may not increase property taxes. If carefully installed, it can later be dismantled and moved to a new location without permanently scarring the house.

There are, of course, some prefabs that are permanent and are designed with some regard for the realities of a Northern climate. All these models replace inefficient overhead glazing with a fully insulated roof, while some offer solid, insulated end walls, and others feature laminated wood trusses instead of metal window frames. These models meet most of the criteria for a thermally effective Northern sunwing, but they are still prefabs, designed to be attached to any house, not to one specific house. "In the beginning, we thought of developing a kit too," admits Bob Argue of Sun Shelters. "But no kit can address all situations because no two attachments are alike. Additions are too unique for kits to work well."

Nevertheless, for some homeowners, the appeal of kits is the particular "look" they add to a house. The elegantly curved eaves of Lord and Burnham or the graceful arch of Brady and Sun laminated wood trusses create an aesthetic effect that is not easily achieved by an owner-builder. But style and grace are never cheap — one

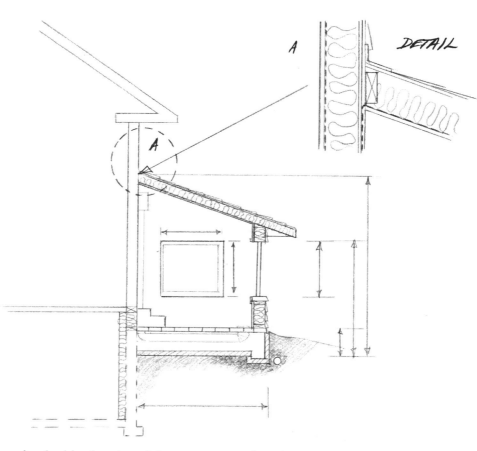

A   DETAIL

popular double-glazed model costs over $100 a square foot, plus shipping, installation and foundation. "If you calculate all the costs, you will find you can buy a custom-designed and custom-built addition for the same price as a prefab or less," says Argue.

Before buying a prefab, the homeowner should follow all the steps set out in the preceding chapters. First, assess personal priorities, and analyze the design to determine if it will perform as desired, then do a full cost estimate, including the price of the kit and such extras as flashing, exterior finishes, interior finishes, shade controls and shipping costs. Get quotes on a

foundation and floor, as well as wiring, plumbing, ventilation and heating. Many manufacturers recommend that their kits be installed by professionals, so find out if there are reliable and experienced installers in the area. "There have been a lot of cowboys in this business over the years, so the consumer should be cautious," warns Michael Glover. When all the costs are determined, divide the total by the square footage of the proposed addition, then shop around. Prefabs have a tradition of style and elegance that dates back to the Crystal Palace, but a custom sunwing will probably complement the style of the house better, be cheaper to build and

PRELIMINARY COST ESTIMATE

Items	Description	Quantity	Cost
Design services			
Demolition			
Foundation			
Building shell			
Doors & windows			
Electricals			
Plumbing			
Heat-loss controls			
Solar controls			
Backup heat			
Subtotal			
25% contingencies			
Total			
Repairs to common wall			
Furnishings			
Landscaping			
Total			
Operating costs:			
Backup heat			
Property tax increase			
Insurance increase			
Hydro			
Total			

more comfortable and less expensive to live in.

PAYMENTS

One of Harry Daemen's priorities in building a sunwing was to keep costs below $500. When all the bills were tallied, he overshot his mark by almost 50 percent, but even at that, his sunwing was about as cheap as they come.

Among the survey respondents, sunwing costs varied from $5 to $85 per square foot. The highest prices were for sunwings finished as living spaces and built by a contractor. In this case, the square-footage cost comes close to the local square-footage cost of new house construction and may even be slightly higher due to the labour-intensive task of attaching new work to an old house that is probably neither plumb nor square. Costs may also be higher than conventional construction because each sunwing is a prototype and, as such, will have to bear the expenses of being a "one-off" project.

If the homeowner builds it, the figure can be cut roughly in half, since construction costs are usually split fairly evenly between materials and labour. Although the homeowner can save some out-of-pocket cash by becoming an owner-builder, the inexperienced homeowner will take longer to do the same job and will probably waste material in the process. The New York State Energy Research and Development survey showed that although most of the respondents installed their own prefab greenhouses, those who hired out the job were more satisfied with their sunwings in the end. "Building it yourself can save dollars but usually costs a lot more of your time and effort than expected," says Don Roscoe. "If people decide to build it themselves, they should be aware that they are doing so for the experience or for the control over the quality of construction, not just to save money."

Material costs can be kept low with good planning and careful buying. A slight adjustment to the sunwing size might save significantly on materials. If the addition is borderline, shrinking it by a few inches may mean using smaller — and cheaper — lumber. (This service alone may justify the cost of a professional designer.) Furthermore, the inexperienced handyman often ignores standard sizing when drawing up plans, then pays a high price for odd-sized materials. Because of the relatively small amount of materials required for a sunwing (as opposed to a house), it may be practical, if the homeowner is hiring out the work, to combine some of the design elements. For instance, instead of hiring a carpenter to build the kneewall and a block layer for the foundation, it may be cheaper simply to extend the foundation above grade so it becomes the kneewall.

Energy-efficient features may add extra material costs. A certain level of low-energy construction is required for comfort, to keep the addition cozy and draught-free, but at some point, the reduction in heat loss may not be worth the cost of materials to insulate. The point at which economics take over from comfort will vary with the use of the space and with the region, since it depends on the cost of backup heat and on the climate against which the space is insulated. In Alberta and Saskatchewan, where there is cheap natural gas, and in Quebec, where there is cheap hydroelectricity, it may not make economic sense to spend a lot on insulation.

Scavenging materials may further reduce the cost per square foot, though the homeowner should be cautious.

Whether it is prefabricated or custom-built, a carefully selected sunwing can add substantially to the value and "curb appeal" of any house.

Used materials may save money at the outset but be more expensive in the long run if the sunwing ages prematurely. Sometimes window manufacturers sell odd-sized windows at a cut rate, or building contractors unload the ends of stock from a large job cheaply. Deals like this can save a lot of money, though the design may have to be reworked to accommodate new sizes. When Aaron Schneider added a sunspace to his Cape Breton farmhouse, he saved hundreds of dollars by redrawing the design to accommodate thermopane glazing from a dealer's clearance sale.

The only way to find out if the rough plans match the pocketbook is to prepare a **preliminary cost estimate**. A detailed costing that puts an actual price tag on the job can wait until the design is finalized and the materials and processes selected. Use the worksheet opposite as a guide to preliminary costing, obtaining ballpark prices from building

suppliers and tradesmen. This initial budget should cover not only actual construction but design, drafting, other professional services, furnishing the new space and the inevitable extras that crop up along the way. The cost of operating the space will also have to be considered, since this will have an ongoing effect on family finances. Err on the side of generosity — it is better to have money in the bank at the end of the project than to live with exposed insulation for two years until the funds are available for plaster and paint.

Do not underestimate the "avalanche factor" associated with any kind of renovation. As Eric Darwin says, "What's $20,000 for a sunwing if you don't have a nice spot to sit or garden? Plan for the finishing touches, not just the room." Those extras can quickly jack up original construction cost estimates. For example, adding a sunwing

breakfast nook means replacing the drywall on the kitchen side of the common wall, which leads to a new coat of paint for the old kitchen, which makes the curtains and countertop look a little shabby, and so on.

The addition may force an upgrading, such as changing 60-amp. electrical service to 100-amp. Halfway through construction, homeowners may suddenly realize that their hard-earned solar gain is going to filter out through the thermal cracks in their house, so they embark on an unplanned and costly foray into reinsulating, caulking and weatherstripping the existing structure. Doing a preliminary cost estimate should help homeowners define the scope of the project: it should include only those improvements that are cheaper to do while the addition is being built, such as sealing air leaks and moisture-proofing the house foundation while it is exposed.

Try to set firm limits before construction begins, and stick to them. Because there are things that cannot be planned for in advance, like discovering wood rot in the common-wall framing, a 20 to 25 percent contingency fund should be added to cover the unforeseen costs that are part and parcel of any renovation.

In terms of operating costs, the space will put more than one strain on the family budget. Aside from increased heat and hydro, property taxes and insurance premiums will also rise with the improvements to the house. A municipal assessor may be able to estimate the extension's effect on the assessed value and property taxes. Calculate these in advance to see if the addition can be comfortably carried within current family finances.

When all the costs are added up, the sunwing may not be within the homeowner's means. Everyone has personal yardsticks for determining the most they will spend on major purchases such as this. In some cases, the condition of the house only warrants a limited amount of effort or expense. Even if the house is physically up to the extension, the neighbourhood may not be: an addition may raise the house value above the local average, making it difficult to resell. If the house may be sold in the foreseeable future or if it was purchased primarily as an investment, it does not make financial sense to put more dollars into it than will be forthcoming at the time of resale.

Though few people are this level-headed (or cold-hearted) about their own homes, chances are that a passive solar addition will pay for itself. Like kitchens and bathrooms, sunwings sell houses. In real estate jargon, they give a house "curb appeal." A 1983 Energy, Mines and Resources report states that "the marketability of housing was greatly

Jim Merrithew

enhanced by solar lighting of clerestory windows, south-facing glazed areas and solar greenhouses, even when these features detracted from the energy efficiency of the housing envelope." The real estate and banking community reinforces the government's finding. When surveyed by *Remodeling World*, they estimated that an investment in passive solar greenhouses is likely to return 100 percent of its value when the house is refinanced or sold. It might even do better than that. Eric Darwin's addition cost a hefty $20,000, but it increased the appraised value of his Ottawa house by $30,000.

Regardless of how the expense is justified, an addition still must be paid for in cash. In most cases, it will be the family bank balance or credit rating, not the resale value of the house, that brings sunwing fantasies back to earth. Family cash flow will also affect how the sunwing construction is scheduled. Darwin found that he had enough cash up front to pay a contractor to frame and close in the addition; then, out of weekly paycheques, he gradually purchased finishing materials, working on the project on evenings and weekends.

If the preliminary estimate indicates that the sunwing will cost more than the family can afford, there are several options. Homeowners can borrow the extra funds, try to get a government home-improvement or small-scale energy-demonstration grant or build the sunwing in stages, pouring the slab and using it as a patio the first summer, closing it in with framing and plastic the second year and, finally, adding permanent glazing and mechanical systems the third. Or the original design can be scaled down. In doing this, it is important to eliminate only those parts of the design that are honestly peripheral and not to compromise quality. In the

long run, it is better to build a good-quality sunwing in stages than to erect a small, inferior sunwing that will have to be replaced in a few years.

Some homeowners try to justify the cost by balancing the initial construction bill against heat savings generated by the passive solar space. There are complicated formulae for calculating payback and life-cycle costs, but as Ottawa designer Michael Lambert bluntly points out, "A person can never justify the expense of building a passive solar addition on energy savings alone." Indeed, the simple payback on Ryan

Campbell's Winnipeg sunwing was calculated to be 100 years!

"I could almost pay my oil heating bills with the interest on the cost of the addition," admits Irene Shumada, "but the sunwing gives me more self-sufficiency and more living space and contributes to my quality of life."

Such benefits cannot be costed. In the end, operating expenses and even construction costs come down to a personal value judgment: how much is it worth to be able to walk into a bright, warm, plant-filled room when the rest of the country is swathed in snow?

Robert Tinker

Building the sunwing in stages can ease the financial burden of construction. Finishes such as exterior siding or plumbing fixtures can often be delayed for a year until they are more affordable. Dick Evers, Winnipeg, Manitoba

6 A Solar Sampler

Seven finished sunwings

'Goodness comes out of people who bask in the sun, as it does out of a sweet apple roasted before the fire'

— Charles Dudley Warner

When Kit Coleman bought his red frame bungalow on the southeast tip of Vancouver Island in 1977, the previous owner assumed he was interested only in the picturesque property. "He thought we would bulldoze the 60-year-old house and build something decent," Coleman recalls. "Little did he know that through my rose-tinted, conservationist spectacles, I was seeing a cozy, energy-efficient, solar-assisted vintage home."

Such idealism is the catalyst that turns cold, hard theory into glowing reality. Eight years later, his renovation almost complete, Coleman basks in the sun that beams through the homemade windows enclosing his porch-sunwing. In winter, the sunwing's collected heat flows into the house through common-wall casement windows that open and close at the dictates of a Thermofor piston rigged with Meccano parts, fishing line and weights. A thousand dollars, a little hard work and a lot of ingenuity built Coleman a sunwing that fulfilled his vision.

Coleman was only one of dozens of Canadians who obligingly opened their sunwings to public scrutiny, sharing their experiences in designing, building and living in a passive solar addition. The seven houses selected as profiles for this chapter cover the geographic and climatic range of this country, from the relatively mild shores of Cape Breton Island to the ice-scoured banks of the Yukon River. Most of Canada's typical housing stock is represented: turn-of-the-century clapboard and stucco farmhouses, the tall brick tenements erected cheek by jowl on narrow city lots during the '20s, the bungalows and split-levels that dominated postwar suburbs and a rejuvenated log house that marries tradition and energy efficiency.

The specific demands of each family, house and site have resulted in seven radically different sunwings. One is a working seasonal greenhouse that is all but inaccessible from the main house, while another is a fully integrated living area with only a few token houseplants. The buffered solar "mudroom" in Nova Scotia is a far cry from the two-storey, plant-filled sun room in central Ontario, though both are attached to renovated farmhouses. And the confines of a city lot in the nation's capital led to a sunwing quite unlike the one that sits on the edge of the uncluttered Prairies. Yet despite such diversity, they share a unifying theme: whether the addition cost $500 or 50 times that much, all of the people who built them are delighted with the new sun in their lives.

These samples are dramatic proof of the flexibility of solar principles. Though each of the seven houses is less than perfect from a solar perspective, the sunwing designs combine energy-efficient construction with passive solar theory so effectively that the additions are thermally self-sufficient. Collected solar heat energy balances the cost of backup heat and in some cases actually lessens the heating load of the house.

Although effective and well-suited to their particular locations, these sunwings are not without problems. Individual circumstances require compromises that are an inevitable part of the design process. As a result, though none of these profiles can serve as a blueprint for another sunwing, together they are a guide to the potential pitfalls and pleasures of sunwing design.

"We accepted the challenge," boasts Max van den Berg, a Manitoban of Dutch/Indonesian descent who longed for a sunny respite from the long, cold Prairie winter. "We proved that with common sense, hard labour and little money, *anyone* can design and build an efficient passive solar addition."

Paul Bailey

DONNELLY SUNWING

LOCATION: Albion Hills, Ontario
DESIGNER: John Hix
BUILDER: Ray Ambraska
SIZE: 12.5 by 30 feet (375 square feet)
COMPLETED: 1983
COST: Approximately $25,000
 ($66 per square foot)
DESIGN TEMPERATURE: minus 3 degrees F
HEATING DEGREE-DAYS: 6,827
MEAN ANNUAL NUMBER OF HOURS OF
 BRIGHT SUNSHINE: 2,047
 (634 October to March)
(Climatic data above for nearest weather
centre: Toronto)
LATITUDE: 43 degrees 52 minutes
 north latitude

Built for under $1,000 using salvaged materials, this Vancouver Island sunwing (previous page) was made by closing in the porch of a renovated bungalow. Well-ventilated in summer, the unheated space is used to shield the front of the house, to hold some hardy houseplants and, primarily, to provide pleasant porch-sitting on clear winter days. Kit and Alice Coleman, Victoria, British Columbia

For years, Gerald and Joan Donnelly talked about expanding their 70-year-old farmhouse with an enlarged porch or "a place for plants." "One day, G.K. came home with *Harrowsmith* and said, 'This is it. This is what I want.'" The magazine profiled John Hix and his award-winning Alpha House, an envelope design that includes a sunwing very much like the one the Donnellys now enjoy.

Aside from a sunny, energy-efficient home for Joan's sprawling collection of plants, the Donnellys were most concerned that the sun porch not "offend the eye. We wanted the addition to look as if it had always been there." And it does. The peak of the existing house runs exactly east-west, so the original roof simply continues its gentle slope to create a two-storey sunwing that covers the entire south face of the structure. Enveloping an uninsulated wall and the single-glazed windows of two bedrooms, living room and dining room, the sunwing acts as an effective thermal buffer against the strong west winds and snow squalls that winter brings to this part of the country, just north of Toronto.

"The dining room always used to be chilly, and the upstairs bedrooms freezing," recalls Joan. "Now, with the protection of the addition, those rooms are comfortable. On sunny days, if we want some extra warmth, all we have to do is open a window or the French doors, and the heat comes flooding into the house. I'm sure we get more free heat from the sun room than we put into it."

Although it is bright and warm on sunny days, some early-morning solar gain is blocked by towering conifers originally planted at the four corners of a homesteader's log cabin. The Donnellys did not even consider cutting them down: "No amount of extra heat could compensate for the loss of those trees."

Unfortunately, an accurate estimate of the sunwing's winter heating cost is spoiled by the overlapping energy demands of an incubator and brooder for Gerald's prize waterfowl. The Donnellys keep the sunwing above freezing with electric baseboard heaters installed under the south glazing, and on cloudy or cool days, they switch on a radiant electric heater installed above the dining room window so that it shines directly down on the rocking chair where Joan sits to watch the birds or to read a volume of her impressive collection of nature books.

The plants seem to thrive in the cool, buffered climate: the Boston fern is a verdant volcano almost 5 feet in diameter; a donkey-tail cascades from the ceiling, spotted with blossoms; and a poinsettia as wide as a Christmas tree blooms scarlet well into spring. By February, the sunwing is crowded with trays of vegetable and flower seedlings that Joan starts for her large outdoor gardens.

"Anyone who endures a Canadian winter deserves a sunwing," says Hix, who designed the space as a meeting place for people and plants. One of his most pleasing design trademarks is south glazing raised 2 feet above floor level, under which is a bench that is used for sunbathing as well as for plants.

The house windows that used to overlook the front yard now open to the plant-filled sunwing. Although the downstairs rooms are a little darker since the addition was built, by judiciously placing skylights opposite the upstairs windows, Hix preserved direct light and a good view of the woods across the road. The skylights are inset into the sunwing roof, which slopes away from the house on either side of a central second-storey porch where a

Although large confiers on the south make this a less than ideal site, the sunwing still gains more heat than it loses, partly because it shelters the adjacent rooms of the house from strong winter winds.

Well-placed skylights light the upper reaches of this two-storey sunwing and retain the view from the second floor of the house.

Gail Harvey

The insulated crawl space under the sunwing is fitted with a continuous polyethylene air/vapour barrier. A layer of sand on the floor absorbs heat blown through vents from the ceiling, then releases it at night to rise through gaps in the 2-by-4 floor.

handcrafted wrought-iron railing that spells DONNYWEIR FARM encircles an outdoor sleeping space the Donnellys refused to relinquish. Though attaching the sunwing was more complex, and careful detailing was needed to prevent leaking, preserving the porch has some unexpected benefits. The flat ceiling in the centre of the sunwing between the two sloping sections dampens noise and creates an air of intimacy without sacrificing the drama of a two-storey space. And the "wings" that now nestle the upper porch have improved sleeping conditions by breaking the strong west winds.

Two screened ceiling vents, protected in winter with movable insulation, exhaust summer heat to the porch, and the south wall of the sunwing is glazed with three sliding glass patio doors that open in warm weather to give the space the feeling of a screened porch. "We've practically abandoned our outside picnic table," says Gerald.

In summer, there is east-west cross-ventilation from windows in those walls, a point of contention between client and

Gail Harvey

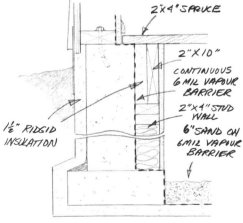

Gail Harvey

architect. From an energy standpoint, Hix disapproves of side-wall windows, but the Donnellys insisted on retaining the view of their swan pond. As a compromise, the Donnellys gave up the east-west sliding patio doors they wanted in exchange for smaller casement windows with insulated weatherstripped shutters that cover the glazing throughout winter.

Aside from the seasonal insulation on east-west windows and upper vents, the Donnellys have no movable insulation to operate. The sunwing does include an inexpensive system to help level off the thermal swings: a fan activated by heat sensors atop vertical ducts in the back corners of the room pumps warm air from the ceiling into an insulated crawl space. There the heat is partially absorbed by a 6-inch layer of sand before the air filters up through slots in the floor.

Inside, the sunwing is finished in light, reflective colours — a grey transparent stain on the floor and woodwork, a light grey paint on the walls. The detailing has been so well done that even from the inside, the sunwing seems like part of the original house.

"There is something especially exciting about having lived in a place for years and years and then having something brand new," says Joan, whose family has owned the house since 1942 and who has no plans to move. "This space changes the house for me, without changing the overall look of the place. It is an outdoor kind of place that can be enjoyed all year round. The dogs understand — on a bright day, this is where they want to be."

ALWARD SUNWING

LOCATION: Mansonville, Quebec
DESIGNER: Ron and Susan Alward
BUILDER: self/David Hansen/
    Tyrone McGowan
SIZE: 13.5 by 9 feet (121.5 square feet)
COMPLETED: 1984
COST: $10,000 materials and labour
    ($79 per square foot)
DESIGN TEMPERATURE: minus 20 degrees F
HEATING DEGREE-DAYS: 8,479
MEAN ANNUAL NUMBER OF HOURS OF
    BRIGHT SUNSHINE: 1,900
LATITUDE: 45 degrees 3 minutes
    north latitude

Ron Alward has worked in the field of solar energy since the late '60s and co-authored a book on attached passive solar greenhouses in the late '70s, yet it was not until 1984 that he finally applied the principles to his own house. "Ever since we bought this place in 1979, I've intended to add a sunwing, but there have been problems with many of the available designs. Most of them looked too much like *additions*: they didn't fit in with the character of the house, largely because of the sloped glazing. Vertical glazing is better suited to most houses, and recent research shows that it works just as well or better from a solar point of view."

Designed as a dining/sitting room extension that would house plants and be energy self-sufficient, the Alward sunwing includes 62 square feet of fixed vertical south glazing. Though it looks like ordinary glass, the windows are Heat Mirror — sealed insulating glass units, with specially coated film stretched between the double glazings to reduce heat loss (see page 122). Alward has worked with researchers from the California manufacturers of the film, and two of his associates in the Memphremagog Group were among the initiators of the original manufacturing company. "I knew the system very well from my own experience and had confidence in it," says Alward. "Besides, we definitely wanted to get away from night insulation for the windows. Our research indicates that enthusiasm for movable insulation fades fairly quickly, even with the keenest of homeowners. So rather than putting in R2 double glazing and blinds that wouldn't be used, we opted for an R5 window."

Window heat loss is important to Alward because his sunwing is fully integrated with the house. A south-facing kitchen window was replaced with an 8-foot opening cut in the common wall, and only a hedge of houseplants separates the sunwing from the rest of the house. "We have many friends who added a buffered sunspace with a door to the house. They thought they would use it a lot, but they don't: it is always too hot or too cold or just not accessible enough. We opened the house to the sunwing because we wanted to make full use of the extra living space and the sun. We are completely satisfied with the result. It has turned into *the* most occupied space in the house."

Because the sunwing is an integral part of the house, its walls are well-insulated. Framed with 2 by 6s encasing 6 inches of fibreglass insulation, the walls are sheathed inside with 1-inch Thermax, which doubles as an air/vapour barrier. Three-quarter-inch lath strapping creates a run for electrical wiring between the air/vapour barrier and the tongue-and-groove pine wall finish.

The sunwing sits on a concrete slab foundation that is 4 inches deep in the centre, thickening to more than a foot at the perimeter, and is insulated underneath with 2 inches of extruded polystyrene. Around the edges, there is 4 inches of extruded polystyrene, which Alward plans to upgrade to 6 inches. Inside, the slab is finished with ceramic tile so the mass can serve as thermal storage. "The tile is a compromise. Dark, unglazed tile would have been best from a thermal point of view, but if the floor is very dark, it gives the room a sombre look, so we chose a lighter brown. Unglazed tile absorbs more solar radiation, but this is used as a dining area, so we opted for glazed tile that is easier to keep clean."

The sunwing itself is only 13 feet long, but the roof stretches across the entire south wall of the house, creating a

The Alwards' south-facing addition is divided into a sunwing on the east and a screened-in porch on the west that may eventually be glazed to extend the solar-heated space. Between the two skylights, the roof slope is flattened to create a balcony off the second floor "solar dormer."

Jim Merrithew

With solar access limited by trees to the east and a porch to the west, vertical glazing and skylights provide sunshine by day and starlight at night.

With extensive experience in the field of solar design, the homeowners were able to draw their own plans.

summer porch on the southwest side. Because the roof shades the kitchen and playroom windows, Alward incorporated two strategically placed skylights made from tempered patio-door replacement units. The porch also shades the sunwing's west window, blocking solar gain after 3 p.m. on winter afternoons. Although the house is oriented 20 degrees east of south, it sits on the southwest slope of a forested hill, and the winter sun does not rise over the treetops to strike the glass until 9:30 a.m. Despite these limitations, the sun is the room's primary heat source. With only one year's experience, Alward is understandably reticent to generalize about thermal performance, but so far, it seems to be functioning as expected: in midwinter, the sunwing often needs the convected heat from the kitchen wood stove and its own electric baseboard heaters to stay warm, but as the days lengthen, it heats itself and some of the house. Overall, Alward anticipates "zero impact" on his annual heating costs.

Although it blocks valuable winter sun, the porch protects the west window from intense afternoon sun in summer.

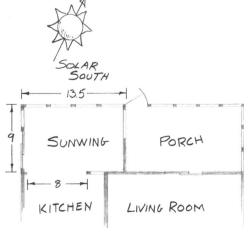

The sunwing is ventilated by triple-glazed patio doors in the east wall and an 8-by-18-inch vent high in the west wall. These, together with other house windows and a south wall overhang, keep the sunwing cool even during the hottest part of the year.

The Alward sunwing is part of a nearly complete general renovation that included reinsulating, reroofing and re-siding the 135-year-old farmhouse. As in most such projects, the new construction had to be adapted to a house that was no longer plumb or square. The south wall of the house leaned distinctly north, so the common wall had to be built out at top and bottom to make it appear vertical. "What started out as a 4 by 4 is now a massive 10-inch pillar," jokes Alward. The floor posed more of a problem. While the sunwing's is perfectly horizontal, the kitchen floor has a 3-inch slope. Again Alward found the solution in illusion. There is a slight step down into the sunwing, and the two floors meet along a 2-by-10-inch pine plate; over its 7-foot length, the plate rises from 1 to 4 inches above the sunwing floor.

Those problems were negligible, however, compared to the benefits the sunwing has brought to the Alward house. "It serves all the functions we originally anticipated and more. It is the prettiest, the brightest and, at night, the most comfy and romantic room. What's more, we now have a good view of the deer that nibble our cedar hedge in winter."

With no common wall to separate the sunwing from the house, the Alwards get maximum use out of the space but must live with some heat drain from the house on winter nights. To improve efficiency and the discomfort of radiant "coolth," they installed Heat Mirror low-emissivity windows. Clay tile over an insulated slab floor helps to buffer temperature extremes.

Jim Merrithew

93

DAEMEN SUNWING

LOCATION: Whycocomagh, Nova Scotia
DESIGNER: self
BUILDER: self
SIZE: 16.5 by 8 feet (132 square feet)
COMPLETED: 1984
COST: $727.11 materials only
 ($5.50 per square foot)
DESIGN TEMPERATURE: minus 1 degree F
HEATING DEGREE-DAYS: 8,049
MEAN ANNUAL NUMBER OF HOURS OF
 BRIGHT SUNSHINE: 1,745
 (567 October to March)
(Climatic data above for nearest weather
centre: Sydney, Nova Scotia.)
LATITUDE: 45 degrees 59 minutes
 north latitude

In 1983, Harry Daemen gave up his engineering job in Ontario and returned with his family to Nova Scotia to enjoy a mid-life retirement. He intended to spend his first year building a passive solar earth-sheltered house, but the turn-of-the-century farmhouse on their rugged Cape Breton acreage was in such good condition that he decided instead to renovate and add a sunwing.

Oriented 10 degrees east of south with no sunblocks on the horizon, the clapboard farmhouse had good solar potential. An alcove in its south face could be closed in to create a bootroom and airlock entry for the kitchen and living room that opened into it. Because the addition would cover two doors and a single-glazed picture window, Daemen expected it would lower his home heating costs, and by adding south glazing, he hoped to use it to start his spring seedlings. At first, he considered raising the existing roof for a two-storey addition, but when he looked carefully at the house, he realized its original lines would be best preserved by matching the new roof to that of the verandah. If the verandah was rebuilt with an east instead of a south entrance, a single path could lead to the sunwing and up a few steps to the verandah. Although these were the "givens" within which his sunwing would evolve, Daemen's actual design was determined by the availability of materials, since his final priority was to keep construction costs to less than $500.

"I didn't quite make it, but I came pretty close," boasts Daemen. "The biggest cost is glazing, but I got mine for $120 from an old hotel — six sets of aluminum storms to make three double-glazed windows. I had already decided that vertical glazing was best overall for heat gain in winter without overheating in summer. Unfortunately, these storms weren't high enough on their own to provide the headroom we needed to preserve the view from the living room, so we elected to put vertical glazing up to the eaves and create a sloped section for plants below."

The sloped glass covers a floor-level growing bed sunk into the south wall foundation and bordered on the outside by an insulated 2-by-4 stud wall. Three feet deep, 18 inches wide and 10 feet long, this in-ground planter has less than 2 feet of headroom but easily accommodates low-growing bushy plants like lettuce, radishes and herbs. "You have to kneel on the floor to garden," admits Daemen, "but isn't that one of the joys of gardening — bending down to work among the vegetables?"

To increase light levels for the plants, the 45-degree slope is single-glazed. Daemen built in a condensation channel to catch moisture dripping down the glass and funnel it back into the growing bed, but in fact, condensation has not been a problem. The glass is kept warm at night with outside shutters made of expanded polystyrene edged with 1 by 3 and glued to plywood on top. Because the underside of the polystyrene is uncovered, it seals well against the glass when the shutter is closed, but when it is opened, the bare foam is sometimes exposed to direct sunlight. Unless the rigid insulation is covered or painted, its life will be shortened by ultraviolet radiation. The shutters are hinged at the top and raised like airplane foils by a manually operated pulley system, with positions for summer shading or winter insolation. "It is no bother to go outside to open the shutters," says Daemen. "I do it on the way to the barn to do morning chores. Of course, they may be targeted for future auto/motor operation if I get really lazy."

There are also shutters on the east wall

For less than $1,000, Harry Daemen closed in an "L" in his Cape Breton farmhouse to create a sunwing that is both a seasonal greenhouse and an airlock entry to the kitchen. The combination of sloped and vertical glass provides light for the plants and winter solar gain without excessive heat loss or summer overheating.

Harry Daemen

glazing — a 4-by-8-foot double insulating glass unit donated by a friend. A vertical 2 by 4 splits the large pane visually, creating proportions similar to the south glazing and accommodating two narrow shutters that are easier to operate and less prone to wind damage than a large one. Hinged on their north edge, they open sideways to a maximum of 45 degrees to act as sunscoops and windbreaks. In summer, they are removed so that the Daemens can see the fields and trees from the kitchen.

Although the problem of the ungainly width of the east window was solved, its added height still bothered Daemen. "Then it occurred to me to build the sloped, angled canopy that now covers the top portion of the east wall. Not only did it improve the appearance by visually joining the front and side lines, but it also created a space for hiding the vents."

These concealed east vents are connected to a hatch in the northeast corner of the sunwing ceiling. The three south windows slide open vertically, letting in summer breezes that exit through the hatch. In winter, surplus heat is moved passively into the house through the living room and kitchen doors. Ducts for active distribution

Harry Daemen

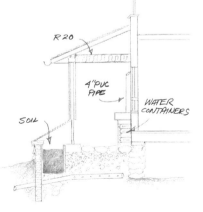

were installed from the ceiling into the basement, but they have proved unnecessary.

Much of the solar gain is absorbed by the mass in the sunwing — a 2-foot rubble foundation insulated on the outside and a flagstone floor of flat stones salvaged from the barn foundation nearby. In its first year, temperatures in the sunwing dipped below freezing only twice — when the outside temperature plunged to minus 4 degrees F, with a windchill factor below minus 40 degrees, after several consecutive days of cloud. Daemen hopes that by fine-tuning the sunwing with water storage, he can keep temperatures above freezing year-round. A planter under the north wall conceals nine 1-gallon water jugs, and an ingenious water/planter holds another 15 to 20 gallons. "I cut off the top quarter of a 40-gallon black plastic drum, plugged the drain holes and inverted it into the bottom of the drum to provide a growing bed above and water storage beneath," says Daemen. "The beauty of the system is that the water warms the soil from underneath, lending itself perfectly to hydroponics."

Daemen does not expect active growth during the depths of winter and therefore provided no backup heat. Common-wall doors are opened only when sunwing temperatures rise above house levels; otherwise, they are kept shut, for the sunwing is first and foremost an airlock entry. "It has been excellent in this respect, making the house much more draught-free and providing a comfortable place to take off wet clothes and boots. The only negative factor about the winter warmth and sunlight has been that my young son uses the planting beds for bulldozing. Oh well — perhaps we'll redefine the second priority as a playroom."

Although it is somewhat inconvenient, Daemen does not mind crouching to garden indoors. At night, the sunken growing bed can be shuttered outside and isolated from the rest of the room on the inside to preserve warmth for the plants.

DUBOIS-DARWIN SUNWING

LOCATION: Ottawa, Ontario
DESIGNER: self/Solarctic Inc.
BUILDER: self/Bollingbrook Inc.
SIZE: 13 by 19 feet (247 square feet)
COMPLETED: 1984
COST: Approximately $20,000
   ($80 per square foot)
DESIGN TEMPERATURE: minus 17 degrees F
HEATING DEGREE-DAYS: 8,735
MEAN ANNUAL NUMBER OF HOURS OF
   BRIGHT SUNSHINE: 2,010
   (647 October to March)
LATITUDE: 45 degrees 25 minutes
   north latitude

At first glance, the downtown Ottawa house seemed to have little solar potential. Though the back of the two-storey brick building faced just east of south, it overlooked a yard completely consumed by a driveway and three-car garage. Rising just 10 feet from the west side of the house was a two-storey brick warehouse that blocked most of the winter afternoon sun. Yet Eric Darwin and Frances Dubois bought the 83-year-old duplex intending to add a sunwing, and in less than three years, their home has become an inner-city oasis of light, warmth and growth.

They began the transformation by tearing down the garage and two unsightly Insulbrick additions built on the house in 1929 and 1949. This created a clear solar window to the south, but the warehouse's shade could only be overcome by thoughtful design. The result is a two-tiered sunwing with ground-level windows to admit morning and early-afternoon sun and clerestory windows that extend solar gain late into the afternoon.

By raising the back half of the sunwing ceiling 4 feet, Darwin not only increased winter sun hours, he also created an efficient summer ventilation system. Warm air rises to the clerestory area where awning windows are left open all summer. A west casement window on the main floor (shuttered in winter) opens in summer to catch the breeze funnelling between the house and warehouse. With good ventilation and a 2-foot overhang to block insolation, the space has never overheated.

The design is the result of more than a year of deliberation. "We did at least 10 variations," recalls Darwin. "We read a lot. We visited people who had sun rooms. We made a list of 'wishes' and 'avoids.' In the end, we hired Solarctic for a one-hour walk through to make sure we weren't missing something obvious, then commissioned them to draw the plans based on our original design."

Solarctic confirmed that the 3-foot poured concrete foundation underneath the demolished additions was sound and only needed to be levelled and insulated on the outside with 2-inch extruded polystyrene protected with pressure-treated plywood. The crawl-space floor was covered with a polyethylene air/vapour barrier, and the sunwing floor joists above were insulated with fibreglass batts, creating a "tempered crawl space" that is unheated but stays above freezing most of the year. Darwin originally ran plumbing through the space but rerouted it after the pipes froze during a January cold snap.

The sunwing walls are framed with 2 by 6s, insulated with fibreglass batts and covered with a continuous 6-mil polyethylene air/vapour barrier over a layer of reflective foil to counteract radiant heat loss. Though Darwin and Dubois hired a contractor to build the weather-tight shell, they finished the sunwing themselves, including the sealed air/vapour barrier. "That is one part of the construction that is worth doing yourself," says Darwin. "No one else will take the time to do it right." Over the air/vapour barrier, the walls were strapped with 2 by 2s, creating a separate raceway for wiring, as well as an extra 2 inches of fibreglass insulation.

The only thermal mass in the sunwing is a double layer of three-quarter-inch drywall on the ceiling and the 14-foot-high exposed brick of the common wall. To isolate the masonry and its airspace from the rest of the house envelope, the bricks at the corners of the common wall were cut out and the addition tied to the inside frame of the house. "From the beginning, we realized no major storage

Despite a narrow city lot and serious shading problems, this Ottawa sunwing is a winter oasis of light and warmth. The two-tiered addition includes clerestory windows that admit shafts of sunshine long after the main floor glazing is shaded by the warehouse wall to the west.

Jim Merrithew

The brick common wall was left exposed to act as thermal storage, and although it helps to counter overheating during warm weather, the brick does not absorb enough heat to keep room temperatures above freezing during a winter night. Originally intended as an unheated, buffered space, the sunwing proved to be such a delight that heating was added so it could be used in any weather.

was possible. We hoped the brick would help, but there isn't enough heat to force into the masonry, so the room cools down quickly on winter nights."

Instead of being stored, excess solar gain flows into the main floor of the house through a window and door and into the second-floor bedroom through a grating. Originally, they planned to use the sunwing as a buffered space, keeping it closed off at night and during cold weather but opening it to the house when it had solar heat to spare. "It cools down so quickly that we found we had to heat it a bit for the plants anyway. And then it seemed silly to have $20,000 worth of space that could be used only three or four hours a day. As we gradually finished off the addition, we found we wanted to use it all the time, whether the sun heated it or not, so it has evolved into a full extension of the house."

A 1,500-watt electric space heater was used to keep temperatures above freezing, but recently, they connected the sunwing into the central heating system and installed Window Quilt insulating blinds to slow heat loss through the sunwing windows at night. "We expected our heating bills to go up some, since we have added many cubic feet of space to the house, and we see any heating contribution from the sunwing as strictly a bonus. Basically, we want to sit in the sun — and we're willing to pay to do it."

Though the quick temperature drop in the sunwing was disappointing, the third tier of the design proved an unexpected delight. The clerestory roof was used as a base for a window seat/plant bench extension to a second-floor bedroom, replacing a former door with a broad bay window incorporating 21 square feet of glazing. The window was engineered to take another bay window or small

Jim Merrithew

balcony above it when the attic floor of the house is eventually renovated.

This triple-tiered design steps the tall narrow house gradually toward the south and ground level, softening its otherwise starkly vertical lines. Because the design itself is so conspicuous, Darwin and Dubois relied on proportion and details to make the new construction compatible with the old. The windows match the style of others in the house, the wooden eaves and soffits mimic the originals, and the wooden pillars, railing, latticework and screen door give the sunwing the overall look of an old-time verandah.

"We paid lots of attention to detail, and having seen many similar additions, I think we really got good value in ours," concludes Darwin. "The afternoon shading from the warehouse still bothers us a bit, even though we knew it would be there, but overall, we got exactly what we wanted — a place to start seedlings, look at the garden and sit in that wonderful sun."

The clerestory windows not only extend winter solar gain, they also "air-condition" the room in summer, as cool air entering through windows on the north side of the house pushes warm air up and out the upper windows. The plants suspended from the ceiling to take advantage of the clerestory light are lowered on pulleys for watering.

To limit heat loss through the airspace that ventilates brick houses, the sunwing was attached to the inside frame wall after removing the corner bricks.

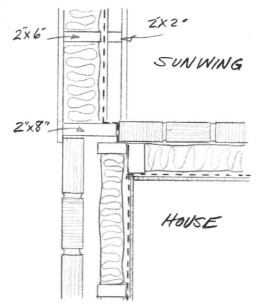

2"x6"
2"x2"
SUNWING
2"x8"
HOUSE

POWELL SUNWING

LOCATION: Priddis, Alberta
DESIGNER: Michael Kerfoot,
    Sunergy Systems
BUILDER: self
SIZE: 32 by 13 feet (416 square feet)
COMPLETED: 1983
COST: $23,000 materials and labour,
    including hot tub ($50 per square foot)
DESIGN TEMPERATURE: minus 27 degrees F
HEATING DEGREE-DAYS: 9,703
MEAN ANNUAL NUMBER OF HOURS OF
    BRIGHT SUNSHINE: 2,199
    (759 October to March)
LATITUDE: 51 degrees 03 minutes
    north latitude

When Greg and Linda Powell built their log house south of Calgary in 1978-1979, they fully intended to fit it with active solar heating panels. "The roof was designed at the right angle for an active system," explains Greg, "but the cost and efficiency of active solar became less attractive, so we decided to build a passive solar collector instead."

Although heating was their original priority, they soon realized that the option they had chosen would also create a very pleasant living space that could, with some compromises, be adapted for growing plants as well. The result is a truly hybrid sunwing that is relatively energy self-sufficient, contributes some heat to the house and accommodates a veritable jungle of houseplants and a few vegetables as well as the family hot tub. "We were using the hot tub for a little R&R between sprees of construction even before the windows were all in or the decking was done," says Linda.

The hot tub is sunk in a cedar deck on the west side of the sunwing, where it is exposed to late-afternoon sun and a spectacular view of the foothills. Two steps down in the middle of the sunwing is an open tiled area that is used for eating and playing, while the east side is devoted entirely to plants. Linda grows salad produce in pots on cedar shelves ringing the south and east windows and in a 4-by-1½-foot wheeled planting box that can be moved with the light. At the back of the sunwing are shelves where she starts garden sets under grow lights.

"I used to grow all sorts of things — eggplant, cucumbers and even cantaloupe," says Linda. "You go through the stage where you have to try these things, but they take up so much space, take so long to produce, and deplete the soil so much that eventually you settle down to what is really useful.

My main thrust now is bedding plants, salad greens, herbs and tomatoes: I bet I only buy tomatoes five times a year."

The Powell sunwing uses no backup heat to keep the room at a comfortable temperature for plants, though there is a furnace duct near the hot tub to keep the water from freezing in an emergency, such as a broken window. "The addition usually stays above 58 degrees F," says Linda. "When the outside temperature goes down to 40 below, the sunwing might drop to 56 degrees, but we never worry about it freezing."

Part of the credit goes to the extensive thermal mass in the addition. Aside from the hot tub, which is well-positioned to absorb late-afternoon solar gain, the 8-inch-thick concrete slab floor sits on 50 cubic yards of fist-sized stones encased in an insulated preserved-wood foundation. Lengths of 5-inch-diameter perforated drainage pipe were sunk into the stones at 2-foot intervals along the north and south perimeter of the sunwing, and a fan blows warm air from the top of the sunwing through a duct into the north wall pipes. The heat disperses naturally through the stone, and cooler air rises up through the south wall pipes. This cools the room during the day and ventilates the plants. At night, the fan is turned off so warm air rises naturally from the rock to keep the sunwing temperature from falling too low.

"The addition heats itself almost entirely," says Greg, "and it helps heat the main house too. The greenhouse floor is 3 feet lower than the main house, so heat rises naturally into the house through a door and a window." However, because the log house has undergone extensive recaulking since the new construction, the effect of the sunwing on house heating is hard to estimate. The Powell's average monthly

Extending across the entire south face of their Alberta log house, the Powells' sunwing is large enough to accommodate the hot tub, a dining area, several plant benches and a convenient potting table.

103

Although the west wall glazing contributes to late-afternoon overheating in summer, it offers extra hours of winter sunshine for lounging in the hot tub.

fuel bill — including cooking and water heating — totals only $25. The sunwing provides virtually all house heat from the end of March to the end of October, and according to Linda, "we would definitely notice a decrease in comfort if the addition wasn't there."

The sunwing walls are entirely glazed on the south, east and west with 4-by-8 wood-frame windows set between 4-by-6 cedar pillars that rest on a 2-by-10 sill plate. The heat loss through this expanse of glass is reduced at night with tracked radiant-reflective curtains made of metallized ripstop nylon, an early design developed by Sunergy Systems. "On a scale of 1 to 10, I'd rate the curtains 7 for aesthetics and 9 for comfort. With all that glass, the minute the sun goes down, you know the curtains are open," says Linda. "I don't find it a chore to operate them. It only takes a minute and has become a natural part of getting up in the morning."

She is less enthusiastic about another winter-morning ritual: mopping up the puddles of condensation that drip onto the windowsills when night temperatures drop below 0 degrees F. Because the hot tub and plants keep humidity levels around 55 percent, the Powells are considering investing in a dehumidifier. In the meantime, Linda spends hours every spring recaulking the windows with silicone and re-Varathaning the sills so that the moisture will not damage the wood or the sealed insulating glass units.

Because the walls are completely glazed and the sunwing is oriented 13 degrees west of south, the room tends to overheat in summer. It is ventilated only by small windows low in the west and low and high in the east walls as well as by a patio door and window that open to the south. "On summer nights, we have to close off the room, or the rest of the house gets too hot," says Greg. Landscaping on the west side of the addition will block some of the hot afternoon sun, but Greg would also like to install automatic venting with a

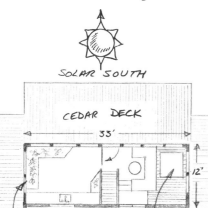

SOLAR SOUTH

CEDAR DECK

33'

12'

PLANT BENCHES

HOT TUB

manual override to flush out the summer heat.

Despite these minor complaints, the Powells rate their sunwing a success, so much so that plans are now underway to build another extension above the first — a second-storey sunlit bedroom. When it is completed, the sharp slope of the original active solar roof will be entirely submerged in the gentle sweep of sunwings.

Roger Vernon

For the Powells, the hot tub is one of the greatest pleasures the sunwing provides, but it is not without its drawbacks. In winter, condensation frosts the windows every night, melting into pools of water that must be mopped up every morning.

The sunwing sits on a preserved wood foundation that is lag-bolted to the concrete foundation of the house. A duct that runs the length of the ceiling carries solar-heated air into the rock bed, thus lowering the peak temperatures in the sunwing. The stored heat gradually radiates through the concrete slab floor into the room, blunting the nighttime lows.

WILKINSON SUNWING

LOCATION: Whitehorse, Yukon Territory
DESIGNER: self/Robert Mason
BUILDER: self/friends
SIZE: 26 by 14 feet (364 square feet)
COMPLETED: 1981
COST: $13,000 materials only
  ($36 per square foot)
DESIGN TEMPERATURE: minus 45 degrees F
HEATING DEGREE-DAYS: 12,550
MEAN ANNUAL NUMBER OF HOURS
  OF BRIGHT SUNSHINE: 1,825
  (448 October to March)
LATITUDE: 60 degrees 43 minutes
  north latitude

The sub-Arctic hardly seems the place for a passive solar greenhouse, yet nowhere would its benefits be more appreciated than in Canada's Far North, where the growing season is short and the heating season long. Wayne Wilkinson's Whitehorse greenhouse is living proof that sunwings can indeed survive north of 60 degrees latitude, cutting both grocery and fuel bills in the land of the midnight sun.

"Up here, you can't put frost-susceptible plants in the garden until July," says Wilkinson, "but by the beginning of February, my greenhouse beds are thawed, and I can start a new growing season. I'm usually harvesting by May, when there is still snow on the ground outside!"

The Wilkinsons grow enough salad vegetables to feed a family of seven and can, freeze or harvest enough tomatoes, cucumbers and peppers to support their craving for equatorial cooking all year round. As well, they sell an estimated $1,000 worth of surplus vegetables and bedding plants a year, most of them to a local "wholesome food" restaurant. His harvests are pure profit, for the greenhouse runs on free solar heat alone. Not only does the sunwing stay above freezing nine months of the year, it collects enough surplus warmth to reduce household heating requirements by 20 percent. In Vancouver, that might be a negligible savings, but in Whitehorse, where the furnace runs year-round, it is significant: the sunwing saves Wilkinson $425 in heating oil each year.

Despite its impressive performance, the Wilkinson sunwing arose from less than ideal conditions. It extends directly from a rear house wall that faces 57 degrees east of solar south. As a result, its northeast side is solidly insulated (R35), but the southwest wall is two-thirds glazed to capture as much afternoon sun as possible before it sinks below a neighbour's house. The back half of the roof is insulated to R62, and the front is double-glazed at a slope of 50 degrees to give the plants optimum light. The south wall is slightly sloped (75 degrees) for maximum solar penetration late in the fall. "At angles this close to vertical, it would be cheaper and easier to simply construct a vertical wall," agrees Wilkinson. "But that would have meant a longer roof, which would have shaded more of the greenhouse and blocked the view from the bedroom windows."

To compensate for the temperature extremes caused by sloped and overhead glazing, the greenhouse-sunwing includes three stages of thermal storage. The first is a thin layer of stucco — the exterior house finish — which covers the common wall. Because of its large surface area, it reacts quickly to sharp rises and falls in temperature. The 63 cubic feet of gravel in the hydroponic growing beds soak up longer-term heat surpluses, and the 247 cubic feet of water that bathe the gravel remove heat for deep storage, returning the warmth at night when the fluid recirculates through the beds.

All three storage systems are strictly passive, but even so, the greenhouse maintains growing temperatures with no backup heat for nine months of the year. In the depths of winter, the addition is simply closed down. "We don't try to extend greenhouse operation beyond what the sun will support," says Wilkinson. "The longest growing season has stretched into early December and the shortest to mid-October. Consistently, the beds are thawed and the greenhouse is maintaining overnight temperatures adequate to sustain plant life by early March."

Wilkinson's desire to have light for his plants and good solar heat collection without blocking the view from his bedroom windows resulted in a triple-sloped design. The lower glazing is tilted at 75 degrees for maximum heat gain, the upper windows are angled at 50 degrees for spring and fall light penetration, and a superinsulated R62 roof minimizes heat loss.

David Arsenault

The hydroponic growing system that produces bumper crops of vegetables helps keep nighttime lows above freezing. The water that bathes the roots of the plants soaks up surplus heat absorbed by the gravel in the planting beds. The warmth is returned to the greenhouse at night when the fluid is recirculated through the growing benches.

This cross-section shows the low, insulated kneewall, the double sloped glazing to admit maximum heat and light and the R62 insulated roof supported by a knee brace. Cool air from the house basement is heated in the sunwing and returns to the house through main-floor windows.

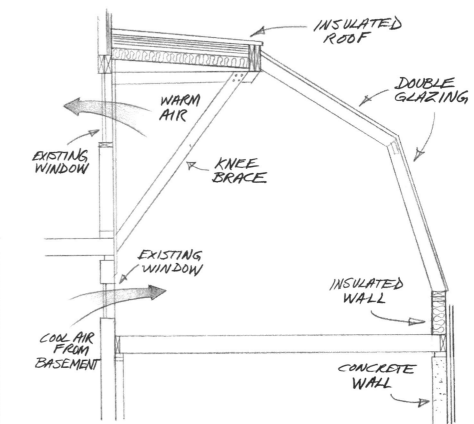

David Arsenault

Much of the time, the greenhouse collects more heat than it needs. The surplus can be vented into the house through operable windows high in the common wall, with sliding basement windows acting as cold air returns. The buffering effect of the greenhouse reduces the heat load on the house year-round and completely eliminates the need for other sources of heat from April to October. Wilkinson's only complaint is that the heat is convected into the back bedrooms adjacent to the common wall, instead of into the front of the house where it would be more useful and better appreciated.

Although he once vented *all* surplus heat into the house, Wilkinson found that at certain times of year, the high humidity caused some condensation problems and aggravated the allergy symptoms of his wife, Donna. Water-vapour levels in the greenhouse range from 18 to 96 percent but are most problematic in late summer when the greenhouse is full of well-developed plants. To lower humidity levels, Wilkinson vents the greenhouse air outside in the morning for an hour or so until it drops below 60 percent, then he opens the common-wall windows.

Condensation in the greenhouse itself has not been a problem, and Wilkinson feels he could have saved $100 by framing the addition with standard construction lumber instead of preserved wood. More significantly, material costs would be $2,000 less had he chosen plastic glazing over double sealed insulating glass units, though that would have sacrificed his mountain view. Furthermore, if the greenhouse had been built on a standard insulated 4-inch concrete slab instead of a full basement, he estimates $4,000 could

Fresh produce north of 60 has repaid the Wilkinson family handsomely for its sunwing project. The sale of vegetables and bedding plants adds $1,000 a year to the family coffers, and the greenhouse cuts the heating bill by hundreds of dollars.

have been saved. He originally planned to build a foot-thick slab floor with buried ducts to circulate peak air through the mass, but because the amount of concrete was exactly what was needed for a basement, he got twice the space for only a small outlay in wooden forms, which were recycled into the greenhouse. The extra labour was his own. Although the basement is insulated on the outside to R15, the concrete has never been used for storage because there is enough thermal capacity without it. Hence, from an energy standpoint, a standard 4-inch insulated slab would have been sufficient.

The sunwing's only main-floor entrance is an exterior door on the northeast wall. To enter the sunwing, Wilkinson climbs a ladder from his basement workshop, which is connected to the house basement. "Access has been a continuing problem," he admits. "I go to the greenhouse every day, and I don't think I ever make the trip without kicking myself for not making a more convenient connection with the house. I haven't wanted to lose any bedroom space at the back of the house, but now I have moved my office there, so I will probably put a door in that wall, with a little balcony extending into the greenhouse. In the meantime, it is incredible how a minor annoyance like that can cause a great deal of dissatisfaction with a project."

Although the sunwing is completely buffered from the house, the whole family manages to enjoy the sun. "At this time of year, in early April, the kids love to play out there. They strip off their clothes and throw water at each other, running around the growing beds. Even though outside it is still winter, inside it feels like summer."

109

DOWLER SUNWING

LOCATION: Russell, Ontario
DESIGNER: self
BUILDER: self
SIZE: 27 by 11 feet (297 square feet)
COMPLETED: 1982
COST: $8,000 materials only
  ($27 per square foot)
DESIGN TEMPERATURE: minus 17 degrees F
HEATING DEGREE-DAYS: 8,735
MEAN ANNUAL NUMBER OF HOURS OF
  BRIGHT SUNSHINE: 2,010
  (647 October to March)
LATITUDE: 45 degrees 17 minutes
  north latitude

On Christmas Day 1983, as Rory Dowler sat in his bright, warm sunwing, munching a sandwich of fresh-picked lettuce and tomato, he noted with some satisfaction that he was living a long-standing dream. With no previous design or construction experience, he had built an addition to his house that was not only a sun-filled, energy-efficient living space but also a "greenhouse" that would keep him supplied with toasted tomato sandwiches all winter long.

"I'm certainly no horticulturist," disclaims Dowler, "but it is amazing how you can put such a relatively small amount of time and effort into something and reap such large rewards. That first year, I had tomato plants 8 feet high in no time!"

Except for the high proportion of glazing, Dowler's sunwing looks like a conventional house addition. Attached to the back of his two-storey brick-and-cedar house, it faces just slightly east of south. Although some trees close to the east and west walls provide summer shade, others have been cut so that the forest edge arcs deeply around the south glazing, creating a clear path for the winter sun.

The sunwing sits on a full foundation that extends to the same level as the house foundation. "For the difference in cost between a full wall and one to the frost line, I decided it wasn't worth the risk of having the foundation shift when it was supporting all that glass. I didn't realize till after it was poured that it should have been pinned to the house foundation, but I have drawn a line at the junction of the two foundation walls, and so far, there hasn't been any movement."

Most of the excavation was filled in to grade level, but a concrete wall 9 feet from the west side creates a basement

potting room under one-third of the sunwing. "That was a really good move," enthuses Dowler. "When you're working with the plants, you can track the dirt downstairs instead of into the house. I keep my pots and extra earth down there too."

The foundation was insulated with 4 inches of extruded polystyrene above grade and 2 inches below, with the full 4 inches extending down to the footings around the basement area. The basement floor is gravel, and the floor above it is cedar decking, so any water spilled in the sunwing just filters through the cracks into the basement and drains harmlessly through the stones. The middle third of the sunwing is floored with interlocking paving bricks, but the eastern third has the same red-cedar decking. "The living room patio doors are two steps up from the floor level of the addition, so I raised the east section with decking," explains Dowler. In doing so, he created a separate sitting-dining area that is a step up from the plant area and a step down from the house.

The foundation wall, backfill and paving bricks all contribute thermal mass to store solar gain, and although the brick wall at the back of the sunwing is just a continuation of the outside house wall, it may also help to temper the indoor climate. At first, Dowler considered creating a thermal break by removing a strip of bricks where the end walls attach, but after cutting a doorway through the foundation common wall, he decided he was not up to chiselling any more masonry. Instead, he lag-bolted the new walls through the bricks to the stud walls behind. "I put piles and piles of caulking at the joint, inside and out. There is a little bit of thermal bridging — the bricks close to the exterior feel slightly cooler — but I don't

From the outside, the Dowler sunwing looks deceptively like a conventional addition, yet the interior is a veritable jungle of flowers and vegetables. A pleasant, energy-efficient living space that is warmed almost exclusively by the heat of the sun, it provides the family with salad greens all winter and is an almost perfect example of a hybrid sunwing.

Jim Merrithew

think it is significant. I figured if there was a real problem, I'd have condensation on the bricks, and that hasn't happened.''

The sunwing has 124 square feet of fixed south glazing, including east- and west-wall patio doors that open in summer for cross-ventilation, and though the upper part of the roof is insulated to R40, the lower half harbours five 22-inch-wide single-glazed acrylic skylights. The middle three are 40 inches long, while the outside pair are 10 inches longer. "I knew the addition would darken the living room and kitchen, which had always been very bright, for they were on the south side of the house. So I installed a longer skylight on each end to get a little more sunshine into these rooms,'' explains Dowler.

Nevertheless, the overhead glazing was installed primarily for the plants. "I didn't want to build a real greenhouse with a great expanse of sloped glass because I wasn't that serious about indoor gardening and I wanted the room to be useful for other family activities. However, I *did* want to grow things, so the skylights are a compromise. In some

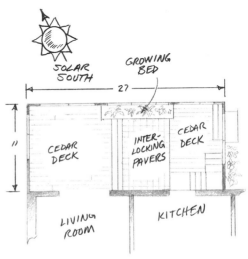

Jim Merrithew

112

A raised deck on the east side of the sunwing creates an inviting nook for basking in the sun as well as a practical transition from the house to the growing area.

ways, I wish I had more overhead glazing, but even with this much, I was worried about overheating." That fear has since proved unfounded: even on the hottest day of the year, it is a very comfortable place to sit.

However, the skylights created an unexpected problem. Throughout the first winter, they continually dripped condensation. Dowler now covers the skylights from November to March with 2-inch extruded polystyrene weighted down with bricks. The snow cover protects the insulation from ultraviolet radiation, and the acrylic stays warm enough that no water vapour condenses inside. Though the low angle of the winter sun combined with snow-reflected radiation partially compensate

for the lack of overhead light, Dowler's plants tend to lean a bit. He hopes to correct that by building a growing bed that will turn on casters to give plants more even light.

Although he solved his overhead condensation problem, frost still appears on the vertical glazing. The humidity problem — caused by the soil and plants — is compounded because the sunwing, like the house, is warmed at night with radiant electric baseboard heaters. Dowler found that a constantly circulating overhead fan and two 17-inch portable fans at floor level move the sunwing air enough to keep condensation to a minimum. He tried installing a trough to catch the condensation, but it quickly overflowed,

so he is now toying with the idea of adding an acrylic slope to the bottoms of the sills to prevent water from pooling on the wood.

Recently, Dowler began monitoring the sunwing to find out how much electric backup heat the sunwing is using to keep nighttime temperatures above freezing. During March 1985, which was unusually cold, he estimates he spent about 15 cents a day. "I easily get that back in daytime heat circulating into the house through the patio doors, so the sunwing must be pretty well self-sufficient. Even if it costs a few dollars to heat in midwinter, it is worth it for the sun and the satisfaction of eating home-grown greens when the ground is frozen solid."

Jim Merrithew

113

7 | Solar Bonds

Joining new and old

'The little things are infinitely the most important.'

— Sherlock Holmes

In a cold climate, houses are constructed around two basic principles: keeping the warmth in and keeping the weather out. Sunwings are no exception, and for the most part, they are put together using standard building practices. Nevertheless, construction of a passive solar addition differs from conventional construction in two important ways.

"Because of the temperature swings and large differences in humidity and air temperature between the inside of the sunwing and outdoors," says Don Roscoe, "the skin of a sunwing has to be more carefully detailed than conventional buildings." The sunwing's dual role as a solar collector and plant-growing space also makes material selection important. As well as having its own distinctive thermal characteristics, the sunwing interrupts the thermal envelope of the house, converting an exterior wall to an interior partition. The effects of these unique aspects of sunwing materials and construction are the subject of this chapter.

Although recommendations are made and details shown, there is no single right way to build a sunwing. House construction is not an exact science, and one builder will pooh-pooh a technique that another stakes his reputation on. Yet all construction detailing aims to do the same things: to keep moisture and cold air from penetrating the building shell from the outside and to keep water vapour and warm air from escaping through the building shell from the inside. The first problem is solved by waterproofing the exterior, "flashing" all exposed horizontal joints so they cannot suck in water, and caulking all cracks so no air can filter in. On the inside, a vapour barrier keeps water out of the wall cavity, and an air barrier, together with insulation, caulking and weatherstripping, keeps warm air contained. Most of the details on the following pages are simply common sense: once a problem is understood, the homeowner can probably come up with an equally workable solution.

COMMON-WALL CONNECTIONS

Because it qualifies as new construction, rather than renovation, building the sunwing is straightforward, but attaching the new addition to the existing house — at the foundation, floor, wall and roof levels — requires special consideration. The care taken and methods used will affect both the structural integrity and thermal performance of the sunwing.

Below grade, the method of attachment depends on the type of **foundation**. If the sunwing has a preserved wood foundation (PWF), one of the easiest to build, simply lag-screw it into expandable lag-shields set in the old construction. If both new and old foundations are poured concrete, drill into the house foundation periodically along the line of attachment, and insert rebar so it protrudes several inches into the forms for the sunwing substructure. If the new foundation is concrete block, then position the rebar so it can be mortared between the joints of the new wall.

However, some experts advise against making a structural connection between the old and new masonry foundations. Although a permanent addition must have a foundation equal to that of the house in both frost resistance and bearing capacity, material shrinkage and the load of walls and roof inevitably cause some settling. Since even minute movements can crack concrete, many builders feel the new foundation should be left unattached to settle

Jurgen Mohr

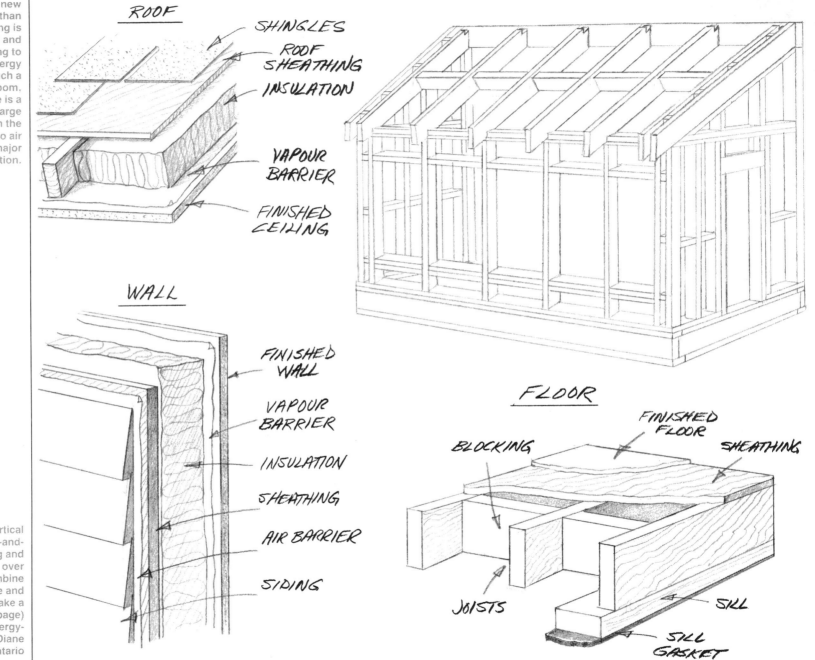

SUNWING STRUCTURE

Because it is new construction, rather than renovation, a sunwing is framed, insulated and finished according to conventional low-energy building practices. In such a highly glazed room, however, there is a disproportionately large number of openings in the building shell, so air infiltration is a major consideration.

Double-glazed vertical windows, tongue-and-groove pine panelling and red quarry-tile flooring over a concrete slab combine with wicker furniture and lots of greenery to make a sunwing (previous page) that is inviting and energy-efficient. Rab and Diane Hunter, Wasi Lake, Ontario

ROOF

SHINGLES
ROOF SHEATHING
INSULATION
VAPOUR BARRIER
FINISHED CEILING

WALL

FINISHED WALL
VAPOUR BARRIER
INSULATION
SHEATHING
AIR BARRIER
SIDING

FLOOR

BLOCKING
FINISHED FLOOR
SHEATHING
JOISTS
SILL
SILL GASKET

independently as a unit, thus avoiding the possibility of damage to the house foundation. Structurally, the sunwing only needs to be attached at the wall and roof levels, where the wood framing easily adapts to any settling.

If the sunwing has a wood **floor**, the joists will likely run perpendicular to the common wall, since this is the shortest span. In that case, lag-screw a ledger (a horizontal board to support the joists) to the common wall. Sill gasket stapled to the wall side of the ledger will absorb any irregularities in the common wall. After the floor is built, frame the end **walls** of the sunwing in sections, then raise them into position. If the exterior siding is brick, attach the new stud walls with lag-screws and expandable lag-shields drilled into the masonry, stapling sill gasket to the back of the stud to tighten the joint. Some designer/builders recommend isolating the interior and exterior masonry of the common wall by removing a vertical strip of bricks so the stud wall can be attached directly to the house framing (see page 101). With a wood-frame common wall, cut away the siding along the line of attachment, and if the plywood sheathing is in good condition, lag-screw the wall through it to the common-wall studs. It may be preferable, however, to strip away the siding, sheathing and building paper, waiting until just before attachment to protect the structural parts of the main building from the weather as long as possible. If there is no stud where the sunwing wall connects, insert horizontal crosspieces between adjacent studs, and lag-screw the new wall to them.

Fill any gaps in the sunwing/common-wall connection with expanding foam insulation, and after the siding is applied, carefully caulk the exterior joint, a chore that may become a regular part of house maintenance. When the

This sunwing roof, though attached at the house eave, matches the slope of a second-floor dormer, thus providing ample south wall window space without spoiling the lines of the house. Rab and Diane Hunter, Wasi Lake, Ontario

Powells built a sunwing addition on their southern Alberta log house, they used platform framing because log was impractical for such short solid walls, but although the sunwing's cedar siding complements the gables in the main structure, it does not move at the same rate as the logs. "Due to the dynamic nature of logs, we have to recaulk every spring," explains Linda Powell.

The sunwing **roof** must be connected structurally to the common wall or roof of the house. If the new roof is a continuation of the existing one, simply remove the fascias and attach the new

rafters to the old. If the sunwing has a gable roof, nail the first rafters directly to the common-wall studs. However, the single-sloped shed roof is the most common variation: secure its rafters with joist hangers or blocking to a ledger plate that is lag-screwed to the common wall or roof of the house.

To prevent leaking, the roof/house joint must be carefully flashed. As a general rule, place the flashing under the high side and over the low side to guide water smoothly over the transition. The most critical area is where two sloping roofs meet, because this is where snow

Jurgen Mohr

117

ROOF CONNECTIONS
The rafters of a shed roof
can be attached either to the
existing rafters or to a ledger
fastened to the house wall or
roof. If the sunwing roof is
gabled, the first rafter is
lag-screwed to the
common wall.

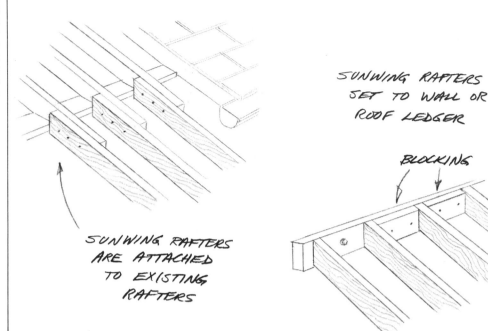

SUNWING RAFTERS
ARE ATTACHED
TO EXISTING
RAFTERS

SUNWING RAFTERS
SET TO WALL OR
ROOF LEDGER

BLOCKING

FIRST RAFTERS ARE
LAG-SCREWED TO THE
COMMON WALL

and ice dams form, damaging the roofing and creating leaks.

Flashing is most easily and effectively installed where a shed roof connects to a common wall finished in wood or aluminum siding. Bend a foot-wide strip of aluminum flashing 4 inches from one edge, then nail the short side under the wall siding, and seal it with roofing compound to prevent rain pushing under the joint. Lay the long side in a bed of roof cement or butyl tape on the sunwing roof shingles. If the common wall is masonry, flashing is a little more difficult: bend a 2-inch lip in the sheet metal, and insert the lip into a slot cut in the mortar joint between courses of brick, then caulk or remortar the joint.

Ventilation is a primary consideration in the roof attachment. Because no air/vapour barrier is perfect,

the roof has to be ventilated to flush out warm air before its water vapour condenses. An airspace between the insulation and roof sheathing, connected with the outdoors, will keep the roof cool in summer and prevent escaping heat from melting snow and creating ice dams in winter. Although there are many roof ventilation systems, the principle is always the same: free airflow between inlets and outlets, with no insulation or framing blocks between.

Ventilation requirements may mean important design changes. The plans for George Dewar's sunwing originally showed 2-by-6 roof rafters on 16-inch centres, stuffed full of fibreglass insulation. Before building officials would approve the plans, however, Dewar had to provide 1 square foot of venting for every 150 square feet of

insulated ceiling area, and a 6-inch clear air passage between the insulation and the roof sheathing. This meant adjusting the design to take 2-by-12 rafters on 24-inch centres, extended beyond the glazing to provide room to install the continuous vents.

A sunwing like Dewar's, attached to the common wall, has a roof ventilation system isolated from the house, with vents at the eaves and the peak or notches at the top of each rafter to provide cross-ventilation. If the sunwing roof is an extension of the existing eaves, however, the ventilation system for the house roof must be considered as well. Some homeowners simply block off the soffits of the house overhang and insulate it to become the sunwing ceiling, inadvertently depriving the house roof of its link with outside air.

When the sunwing and house roofs are continuous, airflow passage must also be continuous, and even then, natural convection may need a boost from fans installed in the house attic.

Air/vapour barrier detailing at the common-wall connection is important because once the sunwing is closed in, the function of that wall changes drastically. Whereas it used to separate the house from the outdoors, it now stands between a uniformly heated space and one that will experience extreme temperature swings. Furthermore, if the sunwing contains a hot tub, pool or a lot of plants, the common wall will separate a high-humidity space from one that is relatively dry, especially in winter.

The illustration on page 121 shows a section of a typical sunwing attachment through a common wall. Note where the air/vapour barrier of the old construction is installed and where the air/vapour barrier of the new construction ends. Trace an imaginary wisp of warm air trying to escape from the sunwing: there is nothing to prevent it from filtering into the common wall and migrating throughout the building shell. Conversely, cold outside air that filters into the insulation cavity of the house can seep into the sunwing. If fans mechanically move air from the sunwing into the house, creating negative pressure in the sunwing, the influx of cold air could be substantial. But this is not the most serious consequence of the air/vapour barrier gap. When warm escaping air migrates into the common wall, it cools down, and the water vapour it carries condenses within the building shell. Not only does this trapped moisture compromise the thermal resistance of the insulation, it can, over time, lead to wood rot in the framing.

The solution is to block off that entry to the building shell with one of the approaches shown on page 121. In the first, the sunwing air/vapour barrier

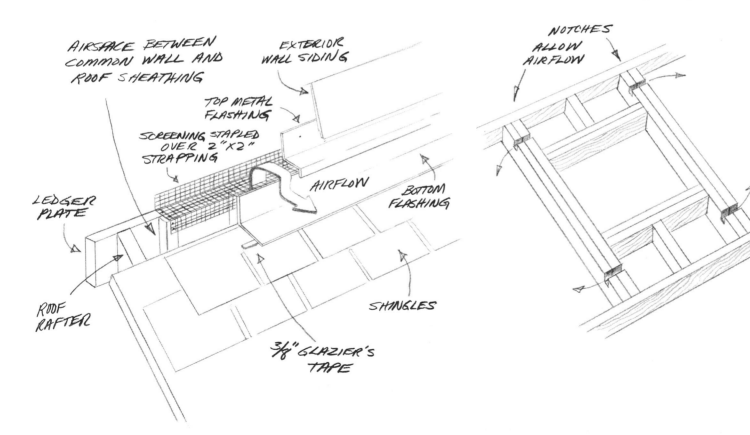

AIRSPACE BETWEEN COMMON WALL AND ROOF SHEATHING

EXTERIOR WALL SIDING

TOP METAL FLASHING

SCREENING STAPLED OVER 2"X2" STRAPPING

LEDGER PLATE

ROOF RAFTER

AIRFLOW

BOTTOM FLASHING

SHINGLES

3/8" GLAZIER'S TAPE

NOTCHES ALLOW AIRFLOW

ROOF VENTILATION
When contractor Don Roscoe connects a sunwing roof to the house wall, he offsets the sheathing slightly to create a 2-inch continuous screened vent that is carefully flashed to provide airflow without water leakage.

If there is a skylight or vent in the roof, cut notches in the rafters so ventilating air can flow around the obstruction.

119

On summer evenings when the sunwing is warmer than the house, these double-glazed sliding patio doors can shut out unwanted heat. Greg and Linda Powell, Priddis, Alberta

is extended through the common wall to connect with the air/vapour barrier of the main house. This approach is most suitable for nonstructural sidings, such as wood and aluminum, that require no special support systems. Connecting the vertical wall air/vapour barriers is relatively simple. The connection at the rafters is somewhat more tedious, since the new polyethylene must be notched around the common-wall studs in order to meet the air/vapour barrier on the other side.

An alternative is to drape the ceiling air/vapour barrier a third of the way down the common wall so the hottest air will not have direct access into the wall cavity. This approach is the simplest solution for brick houses, since removing a couple of courses to extend the air/vapour barrier to the other side can create structural problems.

The *Seymour-Dorgan Report* recommends more drastic measures: insulating the brick veneer on the sunwing side and wrapping it with an air/vapour barrier that is sealed to the house air/vapour barrier only around the perimeter of window and door openings. Although this obliterates potential thermal storage mass, it effectively stops heat loss through the weep holes at the bottom of the wall and thus prevents any condensation forming in the wall cavity. "There is a 1-inch ventilation gap between the brick and the wall," says National Research Council's (NRC) Michael Glover. "It's like leaving a window open all year round. If you are going to heat the sunwing, it is worth going to the trouble of sealing off that airspace, and you can't plug the weep holes without sealing the brick with a vapour barrier."

Openings in the common wall do not require special attention if they connect two consistently heated spaces, but if the

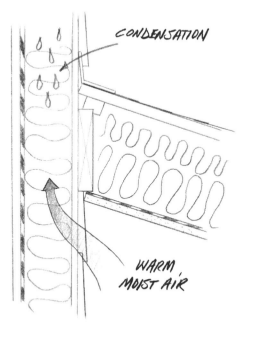

CONDENSATION

WARM, MOIST AIR

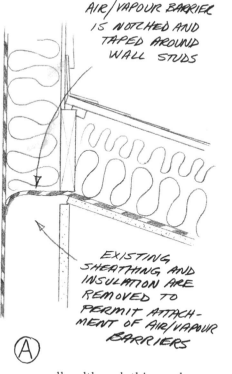

AIR/VAPOUR BARRIER IS NOTCHED AND TAPED AROUND WALL STUDS

EXISTING SHEATHING AND INSULATION ARE REMOVED TO PERMIT ATTACHMENT OF AIR/VAPOUR BARRIERS

Ⓐ

AIR/VAPOUR BARRIER IS EXTENDED DOWN A FEW FEET UNDER WALL FINISH

Ⓑ

COMMON-WALL AIR/VAPOUR BARRIER
If there is no air/vapour barrier on the sunwing side of the common wall, warm, moist air will infiltrate the wall and condense inside the building shell. To prevent this, connect the sunwing air/vapour barrier to that of the house (A), or extend it a few feet down the common wall to foil the escape of rising solar-heated air (B).

sunwing will be closed off during the heating season or if its temperature will drop considerably below that of the house, the doors, windows and heat distribution ports will have to be tightly weatherstripped to prevent heat drain from the house. Before cutting any openings in the shared wall, open a small section to find out if any wiring, plumbing or ducts run through the space. Determine whether or not the wall supports the roof or a second storey: if the common wall is a structural part of the building shell, slicing it open could bring down the house. New openings can be cut in bearing walls but only if new beams are installed to take the load, and even nonbearing walls will have to be braced temporarily while new headers are installed to bridge the gap. Large openings can even be cut in

masonry walls, although this can be a tricky proposition that involves bracing the wall and using temporary jacks to hold it up while installing the new lintel. The danger is that too much or too little pressure can buckle or collapse the wall. If unsure, consult a structural engineer or architect before making an incision in the common wall.

THE GLAZING SYSTEM

Glazing, more than anything else, distinguishes a sunwing from an ordinary addition. The outside "look" of the sunwing, the inside "feel" of the space, its effectiveness as a solar collector and its overall energy efficiency all depend on the choice of glazing.

More than just something the sun shines through, glazing is a complete system that includes the glazing

material, the frame that supports it, and the gasketing and sealing materials that cushion the glazing and seal the frame and glazing into a weather-resistant unit. Because these components work interdependently, they must be both structurally and chemically compatible. If a glazing material expands in hot sunshine and the frames do not flex, the glazing will crack. By the same token, the wrong sealant can eat holes in plastic panels. The components can be bought as a ready-made unit — a window — but if they are purchased separately, as is usually the case for the fixed glazing of the south wall, they must be evaluated as part of an overall system.

Because it has the greatest impact on the thermal and aesthetic success of the sunwing, select the glazing material first, then choose the framing and sealing to

Generic Name	Workability	Commercially Available Form	Trade Names	Thermal Expansion Relative to Glass	% Solar Transmission	Clarity	R Value	Life Expectancy	Comments
Glass	Difficult to cut; installed with putty or glazing compound			1					
		1/8" single-lite float glass	many		84	clear	.9	indefinite	May be used to build on-site vented double glazing
		1/8" low-iron glass	Solakleer		90	trans-lucent	.9	indefinite	Available in double- & triple-glazed tempered units
		Sealed insulating glass unit (SIGU), double, 1/2" airspace	many		71	clear	1.7	indefinite	Patio-door replacement units are the cheapest tempered SIGU.
		Sealed insulating glass unit, triple, two 1/2" airspaces	many		59	clear	2.5	indefinite	Very heavy
		Sealed insulating glass unit, 2 glass with coated polyester film, two 1/2" airspaces	Heat Mirror		53	clear	4.3	film 20 years	
		Sealed insulating glass unit, 2 glass with polyester film, two 1/2" airspaces	SunGain		66	clear	2.7	film 20 years	
		Sealed insulating glass unit, double, one lite coated, 1/2" airspace	Sunglas Low-E Sungate 100 Repla E +		56	clear	2.8	coating 20 years	Relatively new techology
Poly-carbonate	Easy to cut; can be drilled & screwed			8.8					Some haze after a few years; harder & stronger than acrylic; very high impact resistance; transmittance may decrease ± 1% per year
		1/10" monolithic sheet	Lexan		89	clear	.9	3-17	
		1/4" extruded cored double wall	Polygal		77	limited	1.6	3-17	
		1/2" extruded cored triple wall			81	limited	2.8	3-17	
Acrylic	Easy to cut; can be drilled & screwed			8.2					Scratches easily; least yellowing with time; transmittance decreases ± 3% in 10 years
		1/8" monolithic sheet	Plexiglass		92	clear	.9	20	
		1/4" extruded cored double wall	Acrylite-SDP Exolite		74	limited	1.7	20	
Polyester	Easy to cut; fixed with nails or screws with rubber plastic washers			3.2					Yellows with age; can be resurfaced with clear resin when fuzziness reduces transparency (10-12 years)
		FRP - glass fibre reinforced polyester sheet 1 mm thick	Kalwall Filon		88	trans-lucent	.9	13-20	
		FRP - stressed skin panel with aluminum frame 2½" airgap			77	trans-lucent	2.4	13-20	

match. Glazing materials fall roughly into two categories – glass and plastics – and because of their chemical compositions, they have fundamentally different characteristics. To further complicate matters, glass and the three distinct types of plastics are available commercially in a variety of forms, each with its own strengths and weaknesses. The chart opposite attempts to make some sense of the glazing jungle. Since new products are constantly being developed, the glazing material under consideration may not be listed, but the manufacturer should be able to supply information to fill in the blanks.

Since the glazing system is the sunwing's passive solar collector, it is important to know how much solar energy the material transmits; but the flip side of solar gain is thermal drain through the glazing, and materials vary widely in their ability to stop conductive and radiant losses. Both glass and plastics owe most of their heat resistance to the film of still air on their surfaces, so heat loss is generally combated by adding layers of glazing. The trade-off is that each layer further reduces the percentage of heat and light waves able to penetrate the glazing. Light gain may take precedence in a greenhouse-sunwing, but heat loss is more important in solariums.

Although many plastics are clear in single sheets, they become translucent when doubled. Translucency does not reduce the total amount of light admitted – it just scatters the rays, and the diffused light that results is preferred by plants because there are fewer shadows and the light is more even. A solarium with translucent glazing does not feel as hot because direct rays do not touch the skin. In solarium-sunwings, translucency might be favoured by homeowners who want privacy for their

fun in the sun, but it is a definite drawback if the sunwing is intended as a visual link with the outdoors.

Aside from thermal and optical characteristics, the materials should be compared from a structural point of view. Heavy glazing requires more massive framing, and though it might not matter in a solarium-sunwing, in a greenhouse, those supports cast long shadows and block valuable sunlight. A heavy glazing is not necessarily a strong one: homeowners should look at how resistant the material is to acts of man and God, especially if it is installed on a slope. If the glazing is exposed to hail, sprayed gravel, baseballs, icicles or falling apples, it should be able to resist their impact. When a neighbourhood boy heaved a fist-sized rock at the 16-foot glazing on George Matthews' sunwing, the Dartmouth, Nova Scotia, owner-builder was glad he had chosen acrylic instead of glass.

The impact resistance of a material will affect how long it lasts, but so will its chemical composition. Glass is the hands-down favourite in this category, although there is wide variance among the plastics in their vulnerability to moisture and ultraviolet degradation. Unlike glass, plastics are not impermeable to moisture, and unless double-glazed installations are vented, water vapour will gradually migrate through the panel and fog the airspace. Matthews' house was built with skylights made of triple-glazed glass insulating units with an acrylic fourth glazing sealed on the outside. A few years later, the acrylic fogged with moisture that condensed between the glass and plastic glazings.

The southern orientation of sunwing glazing exposes it to considerable daily temperature fluctuations, particularly in winter, when sun-drenched days are followed by frigid nights. Under such

Jürgen Mohr

STORING GLASS UNITS
Unless both panes of a
manufactured sealed
insulating glass unit are
equally supported, one
sheet of glass can slip.
breaking the seal.

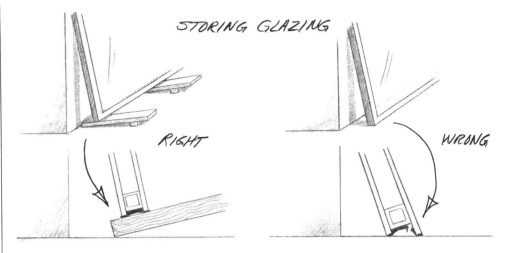

STORING GLAZING

RIGHT

WRONG

conditions, all glazings expand and contract. However, plastics move more than glass, which makes them especially suited to partially stabilized foundations, since they can absorb the little shifts that would crack glass. But it also makes them hard to seal: "When a 14-foot glazing panel moves seven-eighths of an inch over one heating-cooling cycle, it's hard for any caulking to keep up," cautions Roscoe.

All of those characteristics will eventually have to be weighed against the practical limitations of building-code approval, availability and cost — of the glazing itself, the support system and its installation, the operational costs of replacing seals and the overall life expectancy of the product.

Though lacking in high-tech appeal, glass remains the favourite glazing of Canadian architects, designers and homeowners. It is relatively stable, not subject to the expansion and contraction of plastics, and is clear. Glass is unaffected by water or ultraviolet radiation and has an indefinite life expectancy if undisturbed. Its drawbacks are that it is heavy, it is not as easy to work with on site as the plastics,

and it is not an inherently strong material. Untreated, glass has a very poor impact resistance, though when it is heat-strengthened, wired or tempered, it becomes increasingly strong. Sheets of glass over 6 feet long installed on a slope over living areas should be tempered for strength and safety. Because they are tempered, patio-door replacement panels are often recommended for overhead sloped installations.

Although a single sheet of glass transmits a high proportion of solar heat energy — 84 percent, or 90 percent with low-iron glass — its R value, including the inside and outside air films, is less than 1, making it acceptable only for an unheated seasonal sunwing. To improve its heat resistance, glass is usually mounted in multiple lites, and although double glazing is adequate for most of the country, triple glazing is recommended for northern Quebec and Ontario, the Prairies and the Far North. Since much of the heat resistance comes from the air between the sheets of glass, the spacing of the glazings is as important as their number: they should be no less than half an inch apart and can be as much as 3½ inches apart

before there is significant heat loss from convection currents. With sealed insulating glass units, airspaces are usually no wider than five-eighths of an inch because the expansion and contraction of larger amounts of air can break the glass. Unfortunately, multiple glazings greatly reduce light transmission through glass, a definite drawback in greenhouse-sunwings. Low-iron glass offers the best performance on this score, and though expensive, it is worth considering if growing vegetables, since even a minimal increase in light can make a measurable difference in growth.

Multiple glazings can be commercially sandwiched together into sealed insulating glass units, or they can be site-built using single sheets of glass. Although the latter can also be sealed with a desiccant between lites to absorb any moisture in the trapped air, it is hard to make the unit completely airtight. Vented glazing systems are immune to seal failure, since they include small vents that allow outside air to circulate through the airspace. Although condensation may fog the glazing, reducing solar transmission, the moisture causes no structural problems because it can drain to the outside. It can, however, leach salts out of the glass that will gradually scum the inside surfaces, so one of the panes should be removable. Besides being cheaper and longer-lasting, owner-built multiple glazings allow for a combination of materials such as glass and low-iron glass or glass and plastic. They can also reduce conductive heat loss with vinyl or wood spacers, though care must be taken to control air infiltration between the interior lite and the frame.

In commercial sealed insulating glass units, less energy-efficient aluminum spacers separate the panes, and a black sealant holds the unit together. This sealant, unless made of silicone, deteriorates when exposed to ultraviolet radiation and so must be shielded from direct sunlight. Unless all panes in the multiglazed unit are equally supported, the shear action of a slipped pane can break the seal, letting air seep between the panes. This is particularly difficult to prevent if the units are installed on a slope, and in fact, many manufacturers void their standard five-year guarantee under such circumstances. When George Dewar installed 16 double-insulated glass units in his sunwing, neither his contractor, nor the building inspector nor the glass retailer indicated there was any problem with his design, which bared the seal to the sun and left the top lite unsupported. Four of the units have since failed, and the manufacturer refuses to honour the warranty because the units are "improperly installed."

Because multiple sheets of glass are heavy and cut insolation dramatically, some triple- or quadruple-paned insulating glass unit manufacturers now replace the middle lite, or lites, with a sheet of polyester coated with microscopic layers of radiant-selective metallic oxides. In one variation, a high-transmissivity coating increases the amount of solar transmission, and in another, a low-emissivity coating slightly reduces solar transmission but acts as a barrier to room radiation trying to escape. A similar low-emissivity coating is now sprayed on the inside pane of some sealed insulated units, giving double glazing the thermal performance of triple glass without the weight. These windows are still relatively new on the market, they are very expensive, and none have the natural longevity of glass; but it is an area of active research that could result in true "superwindows" within a few years.

On the other end of the financial scale are recycled windows. As the nation's buildings are thermally upgraded, used windows, especially the single-glazed wood-frame variety, are widely available, often at no cost. Recycling these into sunwing glazing involves a lot of labour: transportation, stripping, refinishing, reputtying and perhaps sealing them into multiple glazings. If windows designed for vertical installation are installed on a slope, they will have to be carefully detailed to shed water and snow when tilted. Aside from the work involved, the original framing materials may not be appropriate for the moist conditions of a greenhouse-sunwing. Nevertheless, those drawbacks often pale in comparison to the dollars saved and the satisfaction of salvage. Ross Rabjohn of Hamilton, Ontario, spent three Saturdays dismantling a commercial greenhouse about to be bulldozed to make way for a factory. For $25, he hauled away enough glass, aluminum caps, cypress support bars and steel framing to build a double-glazed, 126-square-foot attached solar greenhouse. "It took a lot of work," admits Rabjohn, "but I'd do it again!"

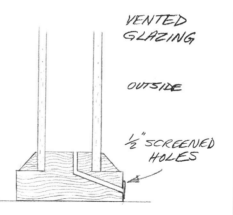

VENTED GLAZING

OUTSIDE

½" SCREENED HOLES

Wood, aluminum and steel are the most common framing materials, though wood is traditionally used with glass. The metals are strong and stable but siphon heat out of the sunwing even faster than the glazing unless the frames are thermally "broken" with a plastic plug between inside and outside surfaces. The biggest advantage of metal framing is that it is thin and casts smaller shadows on plants, but because it is lightweight and is subject to expansion and contraction, it is usually coupled with plastic glazings. Since metal-frame construction requires specialized skills, homeowners who favour that material should consider prefabricated kits, available for both greenhouses and solariums. Many include double glazing and thermally broken frames, but because sealing and gasketing is complex, they will likely have to be installed by a contractor.

For the amateur and professional alike, wood is the preferred framing material. It has good thermal resistance, is easy to work with, is reasonably cost-effective, and it adds an aesthetic dimension that is appreciated in all but the most utilitarian sunwings. It is, however, not as structurally stable as metal. Cracking, warping and rot can be avoided by using dry, good-quality

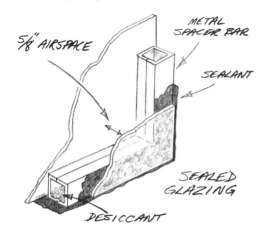

⅝" AIRSPACE

METAL SPACER BAR

SEALANT

SEALED GLAZING

DESICCANT

lumber with straight grain, treating the wood with a preservative or using such naturally rot-resistant species as redwood, cypress or cedar, and by shaping components so that they are less likely to warp and split.

The framing can be nonstructural, supporting only the glazing. In this case, install the framed glazing into a rough opening in the bearing wall, a system suited to post-and-beam construction, which creates a large glazing cavity with few uprights. This was the tack taken by Martin Foss when he built his sunwing near Ottawa. Three vertical 6-by-8-inch posts, one at the centre and each corner, rest on the plate and support a beam of laminated 2 by 8s that in turn carries the rafters. The glazing is installed cheek by jowl in the large south-facing gaps, with only the centre post interrupting a virtually continuous wall of windows. Foss's windows were factory-framed, but they could as easily have been built on site, adding sash, exterior casing and sill to single sheets of glazing or sealed insulating units.

In many cases, though, the frames for the glazing double as the supporting members for the roof. The distance between framing members depends on the windows chosen and affects the size of the lumber required to hold up the roof. Since glass has no structural support itself, it must be held in place by wood that will not split, warp or twist, so use only top-quality lumber for glazing supports. This is an expensive proposition, given the relatively large dimensions of structural frames. Detailing frames that are both structural supports for the glazing material and a bearing wall for the addition's roof can be done several ways, but always install the structural members first — as part of the building shell — then add the glazing. Supported on the bottom by the sill, the glazing can lie against the outside of the uprights or fit between them and rest on glazing stops nailed to the inside of the uprights. Alternatively, it can be seated in a groove rabbetted out of the uprights so that the glazing lies flush with the uprights, but in all cases,

exterior stops hold the glazing firmly in place. As a general rule, it is preferable to use pieces of glass tall enough to avoid the structural and thermal problems caused by adding crosspieces to the framework.

Despite the many different approaches to installation, there are three guiding principles: the system must be structurally secure, it must allow water to drain freely off the glazing and frames, and it must be tightly sealed against air infiltration. Insulated glass units should be seated in the frame on neoprene or rubber setting blocks to accommodate any shifting or settling of the new construction. Seal the glazing to the frame with glazier's tape or an appropriate bedding compound — follow the advice of the glazing manufacturer. This is most critical with plastics, since they require a nonhardening compound that will expand and contract with the glazing. Then seal the glazing frame or the exterior stops to the building shell with caulking to minimize air leakage.

STRUCTURAL GLAZING FRAMES
The wall framing that supports the roof can double as glazing frames if the wood is of good quality and not likely to warp or split. The glazing, held in place with chamfered exterior stops, rests in a rabbetted groove (left) on inside stops (centre) or on the outside of the frame (right).

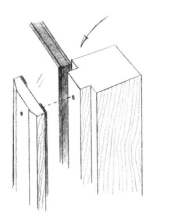

4"x4" RABBETTED TO TAKE GLASS

2"x 6" WITH 1"x2" INTERIOR STOPS

GLAZING SITS ON TOP OF FRAMING SEPARATED BY NAILING STRIP

4"x 4"

Detailing the frames to shed water is more complex; indeed, wood's major drawback is its vulnerability to water. Vertical glazing does not present a problem, but sloped surfaces have to be carefully installed. If cross-members are necessary, keep them as flush with the glazing as possible, with bevelled edges to prevent standing water or ice dams. Cross-ledges can be avoided altogether by using silicone jointing compound between abutting pieces of glass, though professionals usually advise against this technique. If the glazing is sealed insulating units, the siliconed joint has to be covered with flashing anyway to protect the seals from ultraviolet degradation, and unless the glazing is carefully caulked around the sides and bottom, water can seep under the stops and migrate inside.

The glazing system must obviously be impervious to precipitation, but it will also have to cope with inside condensation, particularly in a plant-growing space. If insulated glass units are installed on a slope, the sill will naturally be angled to support both lites and will shed condensation efficiently. However, the bottom sills of other sloped glazings and of vertical installations must be specially detailed to prevent water from pooling. Lower stops should be well caulked and bevelled so that condensation slides down the window, over the stop and onto a sloped bottom sill that has a "drip cut" on its underside to prevent water from dribbling back into the framing. If the growing beds are under the windows, the condensation can drip harmlessly back into the beds; if not, provide a drip channel, an indoor eavestrough, under the sill to carry away condensation. As Bob Argue says, "It is not *getting* wet but *staying* wet that is important in developing rot in wood."

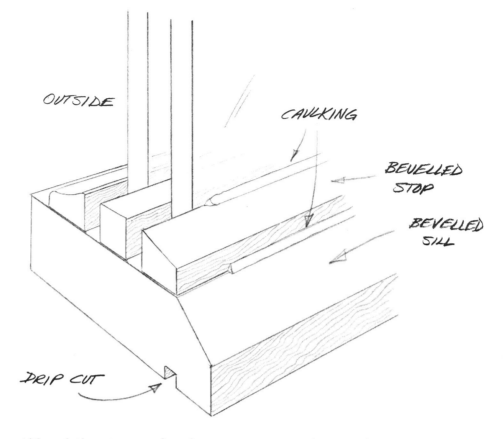

OUTSIDE

CAULKING

BEVELLED STOP

BEVELLED SILL

DRIP CUT

BOTTOM SILL
Unless the stops at the bottom of the window are bevelled and well-caulked, condensation will pool on the sill and damage the wood. A "drip cut" under the sill prevents water from running into the framing.

Although they, too, are sloped, skylights represent a less painful headache than tilted window walls. Installed within the roof framing, they require only nonstructural supports, but they must be carefully fitted to prevent leaking and air infiltration. "If the detailing is done right, there are no leaks," says Hix, who specifies patio-door replacement glass units inset into sloped roofs. "But if the architect or builder has no experience, the client is better off with commercial skylights. Though they are more expensive, they will be covered by a guarantee if they leak." Commercial skylights come in a range of sizes and models; however, the operable types are not recommended unless roof venting is an absolute necessity. Skylights can be flush-

mounted or raised, which interrupts the smooth line of the roof but makes them less likely to leak because a curb guides rainwater around instead of over the glazing. If it is set into the roof framing, be sure the skylight does not block the roof ventilation system: always provide a way for air to move around such obstacles.

Don Roscoe installs skylights and sloped glazing that are guaranteed to leak — to the outside. He rabbets out the 4-by-6-inch rafters so that the glazing sits flush with the outside surface on a bed of butyl glazier's tape, then fastens slate selvage-edged roofing on the outside, over another strip of glazier's tape. Because the tape is not exposed to the elements, it will last a long time, and because it is inset under the selvage

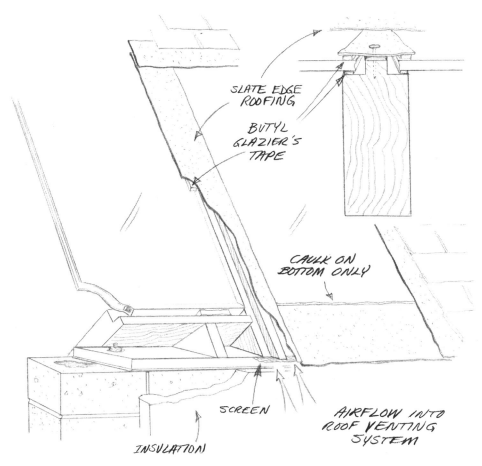

SLOPED GLAZING
INSTALLATION
Don Roscoe uses roll
roofing, glazier's tape and a
minimum of caulking to
install sloped glass that is
fully supported and virtually
leak-proof.

SLATE EDGE
ROOFING

BUTYL
GLAZIER'S
TAPE

CAULK ON
BOTTOM ONLY

SCREEN

AIRFLOW INTO
ROOF VENTING
SYSTEM

INSULATION

roofing, the roll joint thus created is not likely to split − it can move with the expansion and contraction of the glass and roofing. Only the bottom edge is caulked: if water does get under the seal, it simply drains down the rabbet cut and drips harmlessly onto the ground below. He has used this system repeatedly on sloped south walls and skylights with complete success, and even after many years, his clients have reported no problems.

LOW-ENERGY OPTIONS

When building an addition, homeowners can take advantage of the latest construction techniques, but it may not always be in their best interests to do so. In the last few years, new **framing** methods, such as double-wall construction, have increased the amount of insulation in the building shell, and while this is practical for a new house, it may not be suited to a small extension − the heat saved by superinsulating the relatively small wall area of a sunwing is negligible compared to the flood of energy washing out through the windows. Insulate the walls beyond conventional levels only if the house itself is superinsulated or if the sunwing has more solid wall area than glazing.

The roof, however, should be very well insulated because heat rises, and the cold-weather temperature differential between indoors and outdoors will be greater at the ceiling than anywhere else in the building shell.

The type of **insulation** in the walls and roof depends on how the space will be used. Fibreglass batts are most appropriate for solarium-sunwings. If services will run in the insulated cavity, be sure to insulate behind them. Fill the cracks between framing members at windows, doors and corners with loose insulation or expanding foam. Loose fill, either cellulose or fibreglass, can also be used in wall cavities, but it is tricky to detail around wall openings and is more appropriate for floors and ceilings where the cavity is continuous and has less tendency to compress under its own weight.

In greenhouse-sunwings, foam insulation is better than fibreglass because it is inherently moisture-resistant. Leave an airgap between the insulation and the exterior skin, so any moisture that manages to filter through the air/vapour barrier and caulking will not be trapped inside the wall cavity where it can damage the wood.

The greatest source of sunwing heat loss will be its **windows**. The fixed glazings discussed above are relatively energy efficient, as windows go, because they are permanently sealed against outdoor conditions. However, windows, doors and vents regularly open and close, breaking the thermal envelope of the building shell.

As much as possible, such openings should double as part of the summer ventilation system. Donald Hyland of Kingsville, Ontario, originally installed six top-hinged windows in the south wall of his sunwing but finds that two provide all the ventilation he needs, even during Essex County's muggy summers. To cut down unnecessary heat loss, he wisely decided to permanently seal the other four.

Like fixed glazing, operable windows lose heat by conduction through the frame, sash and glass and by infiltration around the joint between the frame and the building shell. To combat conduction losses, windows should be double-glazed with at least a half-inch airspace, and the sash and frames should be made of wood, vinyl or thermally broken metal. A double-glazed sash with an outside storm is a good option if the window will not be sealed with seasonal insulation.

The weakest link in operable windows is the joint between the sash and the frame, where the degree of heat loss depends on the quality of the weatherstripping and the style of the closing mechanism. Windows that latch help pull the sash firmly against its seal, which must be close-fitting and durable so the windows will close tightly even after many years. Foam weatherstripping is initially effective but wears out quickly: wool pile is the best for sliding windows, whereas interlocking or spring types provide a good all-round seal.

Hinged windows, such as casements (side hinge), hoppers (bottom hinge), and awnings (top hinge), generally lose less heat than sliders because they can be closed more tightly. Also, casement windows swing open to "scoop" the breeze, redirecting air flowing along the outside wall into the sunwing, and awnings have the added benefit of letting in wind while keeping out rain. Sliders, on the other hand, are limited to catching what breeze comes their way, and even then, they can only open halfway. The disadvantage of all commercial windows is their square-cut sills, which encourage condensation to pool and seep into the wall cavity. If used in a greenhouse, they should be extremely well sealed with paint, urethane or stain before being exposed to a humid environment. Consider using vinyl or vinyl-coated frames.

The same techniques for thermal efficiency apply to **vents** as to windows. They should be installed with close attention to air infiltration around the

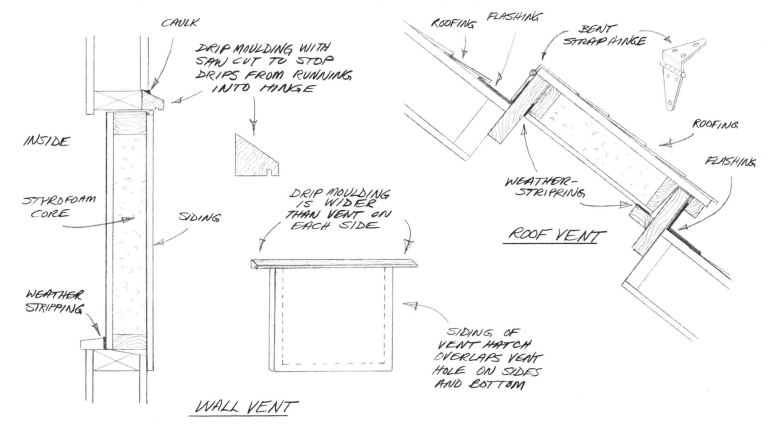

CAULK

DRIP MOULDING WITH SAW CUT TO STOP DRIPS FROM RUNNING INTO HINGE

INSIDE

STYROFOAM CORE

SIDING

DRIP MOULDING IS WIDER THAN VENT ON EACH SIDE

WEATHER STRIPPING

SIDING OF VENT HATCH OVERLAPS VENT HOLE ON SIDES AND BOTTOM

WALL VENT

ROOFING   FLASHING   BENT STRAP HINGE

ROOFING

WEATHER-STRIPPING

FLASHING

ROOF VENT

VENTS
Careful detailing can reduce water leakage and heat loss through openings in the sunwing shell. Vents should be flashed, as shown, well weatherstripped and insulated to the same R value as the wall or roof they penetrate.

**SKYLIGHT OR ROOF
VENT FLASHING**
Openings in the roof are less
prone to leaking if they are
raised above the shingles
with a curb flashing to guide
water around the
obstruction.

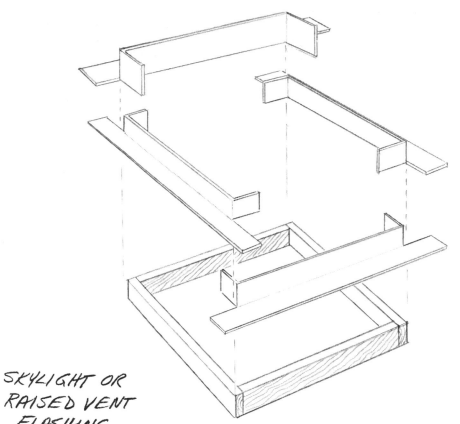

SKYLIGHT OR
RAISED VENT
FLASHING

frame, using a tight-fitting V-shaped weatherstripping instead of a face-sealing type such as foam. Constructed by gluing exterior plywood to a rigid insulation core, they should have an R value equal to the wall they penetrate. Vents can hinge to the outside or inside and from the top, bottom or sides. If they hinge out, they protect the opening from rain but must be screened on the inside and operated from the outside. On the other hand, vents opening to the inside are more convenient, and although louvres can act as a rain shield, the size of the vent must be increased by half to compensate for the reduced airflow.

Roof vents are prone to water seepage, but if they are unavoidable, raise them above the roofline with a curb to direct water around the opening, even though such a tactic breaks the continuity of the insulation. Use movable interior insulation to cover this cavity in the winter.

The same thermal principles that govern windows and vents also apply to exterior **doors** in the sunwing shell. An outside door should be strong, have good resistance to conduction losses and close tightly even after prolonged use. Metal doors are more stable than wood, which tends to shrink and swell with the seasons, making a consistent seal

difficult. However, be sure that metal doors have an adequate thermal break between inside and outside surfaces. Sliding glass doors are often used in sunwings because they combine a large glazed surface with ventilation and outdoor access. Although they are attractive, there is more heat loss with these than with a solid exterior door and fixed glazing. The best place for sliding doors is between the sunwing and the house, where the temperature difference between one side of the door and the other is not as extreme. Like windows, doors should be installed carefully to minimize air leakage behind the jambs and under the sill. Place the lock set with care, because it determines how tightly the door can be pulled against its weatherstripping.

Since exposed concrete has a thermal resistance less than that of a double-glazed window, the **foundation** should also be insulated. If the area inside the foundation will not be heated, 2 inches of extruded polystyrene (blue foamboard) or glass fibreboard is sufficient exterior insulation, but it must be very resistant to water absorption or its insulating value will rapidly deteriorate. Although 2 inches will likely keep the space from freezing, insulation should be doubled within the soil frost zone if the area is to be heated. Exterior insulation is preferred because it seals the mass inside the thermal envelope, protects the parging and prevents moisture and frost from getting to the concrete. Rigid panels can be extended above the critical sill/foundation joint to cover the outside of the framing, creating a seamless armour against heat leaks.

There is some concern within the building trades that replacing plywood or composite sheathing with a moisture-impermeable sheathing such as extruded polystyrene can create a "vapour-locked" wall that traps moisture between the sheathing and the interior air/vapour barrier. In theory, this is not a problem as long as the insulative sheathing is at least 1½ inches thick, is taped at the seams with sheathing tape and is protected with an outside air barrier and a continuous inside air/vapour barrier. "But it is impossible to have a 100 percent perfect vapour barrier," says Roscoe. "You have to assume that some moisture is going to get into the wall, especially from a high-humidity space like a greenhouse. So you have to give it a way to get out of the wall. The outside sheathing has to be able to breathe." Instead of rigid foam insulation, he recommends a permeable insulative sheathing such as Glasclad. These panels are sold with an attached air barrier that lets water vapour out of the wall but does not let rain in. Therefore, it only needs to be taped at the seams to create a continuous shield against cold air infiltration. If plywood sheathing is used, seal the entire exterior with perforated building paper or a spun-bonded polyethylene barrier such as Tyvek or Parsec Airtight White.

Sheathing normally ties the wall framing, floor, joist band and sill together into one unit, strengthening the structure and keeping it plumb and square. Since the insulation panels cannot provide the same structural support as conventional sheathing, the walls may have to be strengthened with diagonal bracing in the framework, plywood sheathing at the corners, or by finishing the walls with panels of plywood siding.

Sturdy horizontal siding like shiplap boards can be nailed through the Glasclad into the studs, but vertical siding, sheet siding and lightweight horizontal sidings like aluminum or

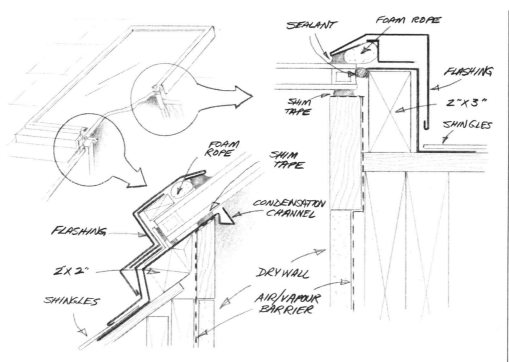

vinyl may leave a wavy surface if installed directly through the easily compressed insulative sheathing. Roscoe's solution is to install the sheathing, tape the edges, then attach horizontal 1-by-4 nailers over the air barrier as a nailing surface for the siding.

Insulation is only half the battle in energy-efficient construction. The high relative humidity associated with plants and the potential infiltration around the perimeter of so many windows make good **air/vapour barrier** installation essential. Six-mil polyethylene should be lapped and sealed with an acoustical sealant such as Tremco to create a continuous envelope inside the sunwing. Detailing around wall and roof openings is especially important.

Extend the sheet over doors, windows, skylights or vents, then cut and seal the polyethylene to the edge of the frame. An even better seal can be achieved by

"bagging" the window or door — attaching a skirt of polyethylene to the frame, then sealing this to the wall air/vapour barrier after the window or door is installed. The only way to ensure that the air/vapour barrier is properly installed (and not torn by the electricians or drywallers) is do-it-yourself construction or close supervision. Infiltration is a major source of heat loss, and wall-cavity condensation problems can be traced directly to sloppy construction practices. The whole point of energy-efficient construction is to force heat to travel to the outside through the insulation, where it will be trapped in the air pockets, and not to bypass it by filtering around vents, wires, doors, flues and wall-floor intersections.

If the sunwing is built over a crawl space — especially one in which the soil is moist enough to grow mushrooms

— it is a good idea to lay a vapour barrier on the earth floor, covered by a thin layer of sand or screed (finishing concrete). Lap and seal the vapour barrier, and extend it up the inside of the foundation wall. If the soil is sandy, good ventilation in the basement or crawl space will probably control humidity, but a layer of polyethylene provides cheap insurance.

Although sometimes recommended for greenhouse-sunwings, an exposed earth **floor** in the addition itself can also create unnecessary and unwelcome humidity. Mary Coyle battled serious condensation problems in her growing space until she covered the bare earth with an insulated floor consisting of vapour barrier, Thermax, sand and cement patio blocks. Not only did the new floor reduce moisture and provide some thermal mass, it was easier to keep clean and eliminated a potential breeding ground for pests.

Some of the energy options in the sunwing will also create special construction considerations. For instance, a floor of 3-inch paving brick laid on wood framing can increase the vertical load on the joists by 30 to 40 pounds per square foot. To support the extra weight, the joists will have to be larger, shorter or closer together. Loads such as water drums that are concentrated over one part of the floor may require special beams supported by bearing posts to distribute the extra load to the footings.

If a concrete slab floor is converted to a hybrid thermal storage mass, it should be thickened to cover the imbedded ducts. Instead of pouring the foundation-wall footing and slab separately, Roscoe saves money by pouring a "raft" that combines the two. The system requires half the formwork of strip footings and can be done in one pour. Concrete or block walls are then built on top, keyed into the raft so they do not slip sideways. To build a raft foundation with imbedded ducts, construct an 8-inch-high perimeter form, and lay 2-inch extruded foam insulation inside, taping the joints. Then cover the insulation with 6-mil lapped-and-sealed polyethylene. Roscoe also drapes the polyethylene over the forms so the wood can be reused as rafters. Lay the ducts on the floor, with outlets fanning to the perimeter from a junction against the common wall. Weigh down each outlet with a small rock, plug the registers with insulation, then secure the ducts against the vapour barrier with plastic tabs — insulation bags work well — over the joints so the ducts will not "float" in the concrete. Shovel a relatively dry mix of concrete on the plastic tabs first to secure the ducts, then pour the floor, being careful not to step on the buried ducts. As a precaution, set markers in the concrete to indicate the position of the ducts, and lift them out before the concrete fully sets.

SERVICES & FINISHES

Services may include wiring to power lights, heaters or fans, plumbing to bring water to plants, and ductwork to distribute warm air either to or from the sunwing.

FOUNDATION-WALL JOINT
To avoid gaps in the building shell, through which warm air can escape and cold air can infiltrate, install a continuous air/vapour barrier and an exterior air barrier. Offset exterior rigid insulation joints and framing joints.

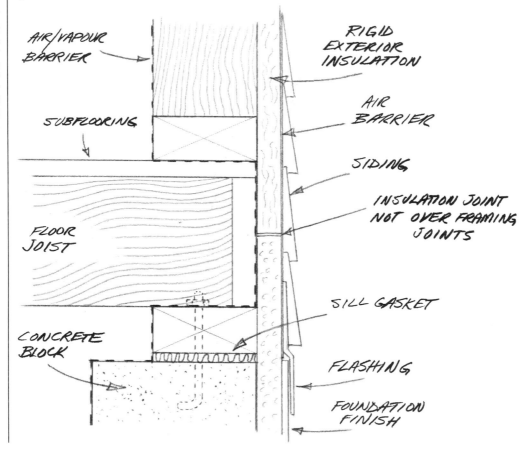

AIR/VAPOUR BARRIER

SUBFLOORING

FLOOR JOIST

CONCRETE BLOCK

RIGID EXTERIOR INSULATION

AIR BARRIER

SIDING

INSULATION JOINT NOT OVER FRAMING JOINTS

SILL GASKET

FLASHING

FOUNDATION FINISH

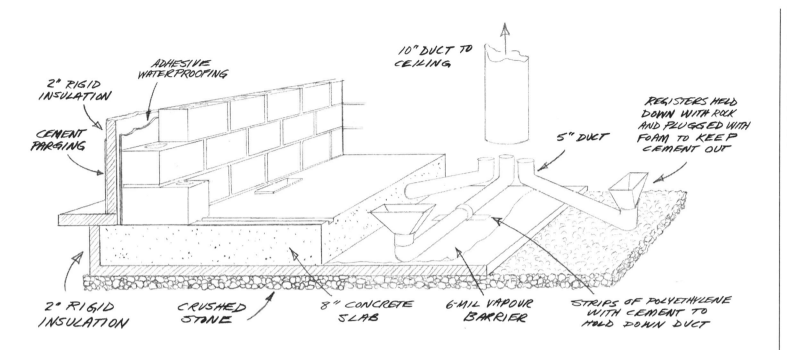

ADHESIVE WATERPROOFING

2" RIGID INSULATION

CEMENT PARGING

10" DUCT TO CEILING

REGISTERS HELD DOWN WITH ROCK AND PLUGGED WITH FOAM TO KEEP CEMENT OUT

5" DUCT

2" RIGID INSULATION

CRUSHED STONE

8" CONCRETE SLAB

6-MIL VAPOUR BARRIER

STRIPS OF POLYETHYLENE WITH CEMENT TO HOLD DOWN DUCT

Electrical needs will vary depending on the function of the sunwing. A greenhouse may be wired for grow lights to stimulate plant growth, propagating equipment to start seedlings, soil-heating cables, fans to counter stratification or actively ventilate the addition and thermostats to control automatic venting or heat-distribution systems. Solarium-sunwings will need duplex receptacles for such electrical appliances as stereos or reading lights, as well as wiring for fans, heaters, thermostats and both inside and outside lighting fixtures.

Select electrical equipment with an eye to energy efficiency. Some fans and motors use less energy than others, and correct sizing is important. Track lighting is preferable to recessed ceiling fixtures, which penetrate the insulated roof and significantly increase heat loss. Since the mechanical and electrical equipment must function in the temperature and humidity extremes of the sunwing, use exterior fittings that are appropriate for damp locations.

The layout of the wiring itself is important. Wherever possible, it should be limited to the common wall so the insulating cavities of the exterior shell are not disturbed. Thermostatic controls must be positioned where they will not be subjected to draughts or direct sunlight, and it is a good idea to think ahead and string wiring for ventilation fans in case passive venting cannot handle the heat load. It is much easier and less expensive to run the wire when the walls are open during construction.

If wiring must be run through exterior walls or the ceiling, provide a special channel for it on the room side of the air/vapour barrier. As Bob Argue says, "This strapping may take extra time and materials, but the electricals are easier to put in, so labour costs should be lower."

It may not balance out exactly in terms of construction cost, but it will in the long run in terms of heat loss." Otherwise, set electrical boxes in a polyethylene "bag" or a polypan — a rigid polyethylene box — sealed to the air/vapour barrier with acoustical sealant or butyl tape.

Wiring for a greenhouse-sunwing must be able to withstand moisture and should be fused with a ground fault interrupter for added safety against electric shock. Consider installing light switches on the house side of the common-wall door so they are not exposed to the extremely moist conditions of the greenhouse.

Plumbing connections will be appreciated in any plant-growing space, whether it is an active greenhouse or primarily a solarium-sunwing. For plants, there should be both hot and cold running water, premixed if desired, since most species prefer room-temperature water and some can be damaged by cold waterings. Plumbing connections — hot and cold water and appropriate drains — may also be required for a hot tub, a partial bathroom or a solar domestic water heater on the sunwing roof.

If the sunwing will be "uncoupled" from the house during the heating season, the plumbing should be fitted with shut-off valves so the pipes can be drained for the winter. In the summer, cold pipes can sweat, so make sure that they are wrapped with insulation or that the condensation will not harm insulation or goods stored under them in the crawl space.

Thoughtful coordination of **finish materials** will not only improve the utility and thermal efficiency of the sunwing but integrate the new construction with the old. Unless the house is relatively new or has recently had a facelift, however, the colour, pattern or style of the original siding and roofing may be difficult to match. New shingles should always be chosen to meld visually with the existing roof, but a contrasting siding, if well chosen, can be very attractive. For instance, the reddish brown brick of Irene Shumada's suburban bungalow is complemented by the cedar-sided sunwing attached to the back. Wood is a popular choice for sunwing siding because it is easy to install and relatively inexpensive, and since the solid surfaces on a sunwing are usually fairly small, maintenance is not intimidating.

Stain is preferred for exterior applications because it soaks more deeply into the wood fibres than paint and sun-rain cycles do not cause blistering. Stain the siding and trim separately so that hidden surfaces are equally protected, but avoid products containing the preservatives pentachlorophenol, chromated or ammoniacal copper arsenate or creosote. They are toxic, and although copper and zinc naphthenate are not quite as dangerous, handle any of them with caution — they preserve wood because they are poisons.

Drywall is the most common interior wall finish for solarium-sunwings, and although it lacks the fine finish of true plaster, it is relatively inexpensive and within the skills of any amateur who has an eye for detail. Because the sunwing is a transition space between indoors and out, natural finishes such as solid or plywood panelling are particularly appropriate. More expensive woods can be used as trim, especially around windows, where it pays to use a good-quality rot-resistant species.

In greenhouse-sunwings, the interior finishes will be exposed to more extreme conditions, particularly high humidity, so exterior siding makes a good inside wall finish. Treat wood or panelling as for outside installations, but avoid using pine or other resinous woods, because the high temperatures will cause the knots to bleed. Alternatively, a special water-resistant drywall is available that can tolerate the humidity of a room filled with soil and plants. Attach interior finishes with galvanized nails, and use construction adhesive as well as nails to fasten bottom window trim.

The walls and ceiling of a growing space can be painted or stained; however, a highly reflective paint to increase light and decrease heat absorption is preferable. A high-gloss white paint specially formulated for

UNPUNCTURED AIR/VAPOUR BARRIER

EXTRA STRAPPING PROVIDES WIRING CHANNEL

CARDBOARD TO AVOID PUNCTURING AIR/VAPOUR BARRIER

134

Jim Merrithew

Perched above the sunwing's clerestory, this window-seat alcove is painted off-white to get excellent light reflectivity without glare. Eric Darwin and Frances Dubois, Ottawa, Ontario

moist conditions − such as an exterior-grade latex or a two-part epoxy paint that cures to a ceramic-like finish − is appropriate. Apply the paint over a primer on absolutely dry wood, preferably before plants and soil are moved in. Because these paints do not breathe, any moisture in the wood will be trapped there, causing the paint to blister and peel.

Solarium-sunwings, without exposed soil beds or banks of plants, do not require such water-resistant finishes, and the reflected light so beneficial to plants would create an uncomfortable glare. A low-reflectivity light grey looks almost white but is not as harsh.

Stain is often used in a solarium-sunwing, and although not as reflective as white paint, it can also be used in greenhouse-sunwings to protect exposed wood. Of the available preservatives, only copper or zinc naphthenate should be used indoors.

The house side of the common wall will already have a finish − though its condition depends on how neatly the openings were cut − but the outside is now an interior wall. Wood, aluminum or vinyl sidings are usually replaced with an appropriate interior finish, and if the sheathing under the siding is uneven, remove it, too, so the new finish can be secured directly to the common-wall studs. If the common wall is brick, it can be covered with drywall or left as is.

8 Hiring Help

Bids and builders

'Experience is the name everyone gives to their mistakes'

— Oscar Wilde

The Daemens, the Przewozniks, the Huttons and the Quirings are typical of sunwing owners across the country who take great pride in having built their own passive solar additions. But a homeowner does not have to be his own carpenter. Once the design is finalized and materials and processes narrowed down, there are other ways to get the project built: hiring a contractor to do the entire job; overseeing the work of skilled tradesmen who do the actual construction; or, like Eric Darwin, hiring someone to erect a weather-tight shell but doing the insulating and finishing oneself.

A contractor hired for the job will take care of all permits, insurance, purchasing and delivery of materials, hiring of subtrades and scheduling. The homeowner who acts as his or her own contractor assumes these headaches, on top of whatever sawing and hammering is not tendered. Contracting involves a lot of running around. It also requires a considerable knowledge of construction practices to keep each phase of construction from being stalled by late deliveries or bad scheduling. To do a good job, the homeowner-contractor must be well-organized and have lots of time, a cool head and enough knowledge about construction practices to get good bids. Doing one's own contracting may not save much money in the long run, since the professional contractor gets a discount on materials, knows which tradesmen do the best and fastest work and presumably is not flustered or frustrated by the inevitable on-site crises. However, for the homeowner with the skill to pull it off, self-contracting offers more control than hiring out the whole job.

A contractor is hired on the basis of his "quote," which includes the price of materials, labour and his own time, but for self-contracting, the homeowner will have to prepare a detailed cost estimate for each stage of construction, listing the kind and quantity of every bit of material that will go into the sunwing — grades, dimensions and species of lumber, quantities and types of fasteners, sizes and styles of doors and windows. List brand names wherever possible. Never let *anything* pass with the rationale, "that won't cost much," and be sure the quantities include what has to be purchased, not what will be used: that half keg of nails or 6 feet of polyethylene cannot be returned for a refund. When the detailed takeoffs are complete, check with suppliers to fill in the cost of each item, remembering to include delivery or shipping charges and fees for rented tools and equipment. Tally the totals, and add 25 percent to cover oversights and surprises. This is the basic material cost of the sunwing.

If the grand total is seriously out of line with available finances, this is the time to go back to the drawing board. Do not wait until the sunwing is half built to find out there is no money to close it in. Major discrepancies may mean redesigning a more modest addition, but if material costs are just a little high, substituting one trim wood for another or settling for drywall instead of knotty-pine panelling may bring the project back within budget. Never skimp where quality is concerned, but trim the frills to the limit of personal tolerance.

Next, prepare a tentative construction schedule, something like the one on page 138. As Don Roscoe points out, "It is important to clearly lay out who is going to do what in a construction project, how long each step should take, and to properly prepare oneself for the process of adding a sunwing." Preparing this tentative schedule will help the homeowner clarify which jobs can

Jurgen Mohr

SAMPLE CONSTRUCTION SCHEDULE

	#Days	Who	COST Materials	Labour
Get building permit				
Stake out sunwing				
Clear site				
Excavate & place gravel	1	Excavator		
Level gravel	1	Self		
Build & level slab form	1	Self		
Prepare for pour: lay drain, ducts, plumbing	1	Self		
Pour slab & polish	1	placer, polisher		
Remove forms, clear site	1	self		
Build foundation wall	1	mason		
Waterproof foundation	1	self		
Insulate & parge	1	self		
Frame walls	2	self		
Sheathe walls	1	self		
Frame roof	1	self		
Sheathe & shingle roof	1	self		
Roof trim, fascia, soffits	1	self		
Install windows & doors	2	self		
Install air barrier	1	self		
Apply siding	2	self		
Trim & paint exterior walls	1	self		
Make vent doors, install	1	self		
Rough in wiring	1	electrician		
Rough in plumbing	1	plumber		
Insulate walls/ceiling	1	self		
Install air/vapour barrier	1	self		
Hang drywall	1	hanger		
Tape & finish	3	taper		
Trim	2	self		
Paint	2	self		
Install electric fixtures	1	electrician		
Install plumbing fixtures	1	plumber		
Finish floor	1	self		
Enjoy				
TOTALS				

realistically be handled alone, which will need the help of a neighbour, friend or brother-in-law and which will be hired out to professionals. Phone calls to a few tradespeople should yield fairly accurate estimates on how much time needs to be allotted for each phase, but be generous — most jobs take longer than expected. Not only will the schedule define anticipated labour costs, but it will also assist in financial planning by identifying the cash needed to pay for each stage of construction.

If a contractor is to build part or all of the sunwing, the homeowner will have to call for tenders on the job, and unless the details of the job are clearly defined, it will be impossible to weigh competitive bids. This is where a good set of plans is invaluable, ensuring that everyone is pricing the same job, and if a specification sheet was not supplied by the architect or designer, the homeowner will have to prepare one. Carefully prepare a detailed written description of construction processes and materials needed to complete the job: if it is not spelled out that the air/vapour barrier joints must be lapped by 6 inches and sealed with acoustical sealant, the homeowner has no right to complain when it is simply stapled to the studs. Although overzealous specifications may scare off some good contractors, it is a risk worth taking, since the homeowner will have to live with the consequence of inaccurate or insufficient specifications for as long as he or she owns the house.

At least three qualified tradespeople, preferably ones with experience in energy-efficient additions, should be asked to bid on a job. Ask friends or the local building supply shop for recommendations, or, as a last resort, try the yellow pages. Make sure each bidder has access to the plans and specifications and submits a written bid on a tender form similar to the one on page 141. The homeowner who has done a detailed costing will already have a pretty good idea what the work should cost, so if a bid comes in higher than expected, ask for a breakdown. Sometimes if a contractor is unfamiliar with a technique, he will quote high to cover himself, but often a simple explanation of what is actually involved will bring the price back in line.

"The homeowner has to do the research and be ready when calling for quotes from contractors who are unfamiliar with the techniques. Basically, it is standard construction except for the vapour barrier, but the detailing is important," emphasizes Bob Argue.

In making the final choice, do not let cost overshadow quality. The lowest bid may come from an inexperienced builder or from one trying to get a share of the local market at any price. Take a look at the person's previous work, and make sure the individual or company is solvent. It takes little more than a pick-up truck and a tool box to call oneself a contractor or carpenter, so the homeowner should be assured that the people hired can and will do the job.

For the protection of everyone concerned, no work should be undertaken without a contract. The contract does not have to be complicated, laced with whereases and to wits, but it must clearly set out the scope of the work, when it will start and be finished, payment and guarantees. The contract usually refers to the drawings and specification sheet so that it is clear both parties are agreeing to the same thing. Either the contract itself or the specification sheets will spell out who is responsible for obtaining and paying for permits and insurances (public

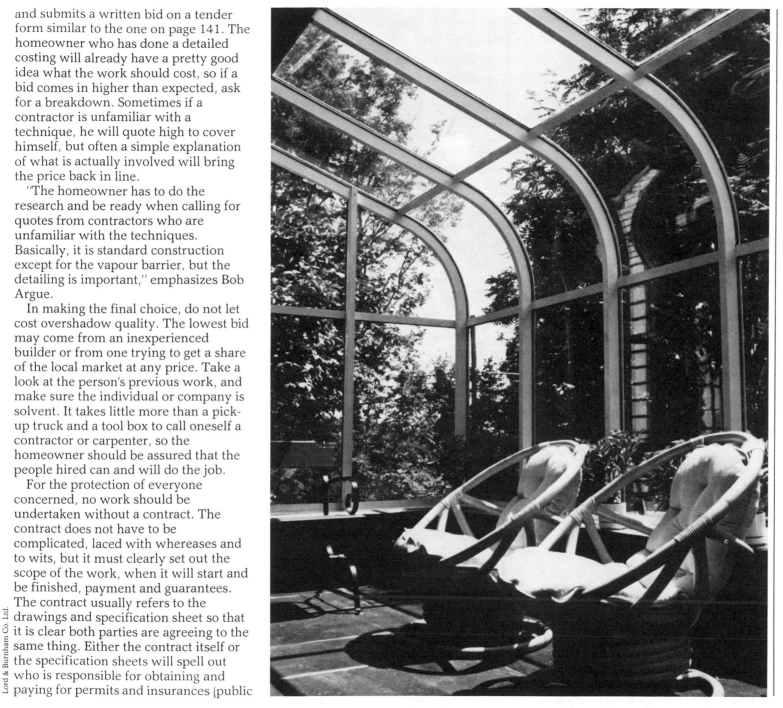

Lord & Burnham Co. Ltd.

For those adding a prefabricated sunwing, serious consideration should be given to hiring a contractor familiar with the brand. Otherwise, inexperienced workers may learn the idiosyncrasies of a prefab at the homeowner's expense.

liability, property damage and coverage for any workers hired).

Clear drawings and detailed specifications do not guarantee that all will run smoothly, however. Martin Foss' contractor built the foundation 2 feet wider than specified, which created a more shallow roof slope. As it turned out, this was a fortuitous mistake because the skylights on order were better suited to the new slope than to the one in the original design. Eric Darwin's house was measured incorrectly, so the windows that arrived on site did not fully glaze the south wall. That problem was also solved amicably when the builder agreed to buy back one of the windows, which was then replaced with two narrow ones. Irene Shumada's addition was originally framed at least a foot too low, impinging on the house windows and producing an unpleasant "shacklike" effect. The framing had to be dismantled and rebuilt, adding a week to the construction schedule. "Be around as much as possible during construction," advises a wiser Shumada. "You are the only one making the financial and emotional investment. Many decisions have to be made during the building process, and once it is completed, it is next to impossible to get poor construction remedied."

This is not to imply, of course, that all contractors are klutzes. Most are honest, responsible tradespeople doing their best to fulfill the terms of the contract. Maintaining a good relationship with the builder of one's dream is not going to be easy, especially when the inevitable human errors, large or small, creep in. Nevertheless, homeowners are likely to get better service if they treat the people they hire with respect. Make priorities clear, but don't tell them how to do their jobs. Be observant so that discrepancies, misunderstandings and unacceptable

workmanship can be dealt with before they become major reconstruction problems.

Regardless of who does the work, construction cannot begin until a

building permit is issued by local officials who have approved the drawings under zoning bylaws and building-code regulations. Inspectors will come to the site periodically to

Not all contractor mistakes are fatal. When the contractor put too shallow a slope on this sunwing roof, the new angle accommodated skylights better than the original design, but less fortuitous mistakes can be costly fiascoes. Clearly written contracts and good communication are the best defence against mistakes. Martin and Betty Foss, Ashton, Ontario

Jim Merrithew

ensure that the addition is actually being constructed according to the plans and to the building code. The code specifies details for every stage of construction – footing sizes, allowable spans for floor joists, roof ventilation ratios – so that all new construction will be sturdy and safe. The building inspector will probably want to see the job after the foundation is built but before backfilling, when the shell is framed but not finished, again after insulation is installed and finally after the project is complete. In a rural municipality, the building inspector may never put in an appearance, which only means the homeowner must be doubly diligent in ensuring the sunwing is structurally sound.

Some provinces have added a section to their building code that deals specifically with renovations, allowing builders more flexibility in meeting the code. In some cases, any changes to the existing building simply have to meet or exceed the standards of that dwelling, but if the local code has no such provision, the new construction, including modifications to the common wall, will have to meet the standards for new construction.

Depending on the jurisdiction, separate electrical and plumbing permits may be required, with inspections after the services are roughed in and again after they are connected. All installations must meet minimum codes, and in some cases, they can only be done by licensed plumbers and electricians. If the homeowner is doing the work, it is essential that copies of the pertinent codes be obtained from the building inspector's office to avoid costly mistakes. If professionals are hired, they will know the codes and abide by them – otherwise, they would not be in business very long.

TENDER FORM

Addition to the residence of _____

_____

at _____

_____

We have examined the drawings, numbered 1 to ____, inclusive, and specifications pages 1 to ____, inclusive, for the addition to the home of

_____

and having visited the proposed site, agree to supply all materials and labour, plant and equipment necessary for the completion of the said addition to the full intent of the drawings and specifications for the sum of ____

_____

We estimate that we can start construction not later than _____ and complete for occupancy by the owner not later than _____.

Signed: _____
For: _____
Date: _____

With permits in hand, the labour lined up, materials ordered and the money either borrowed or saved to pay for it all, the heady process of construction can begin. As the paper-and-pencil plan unfolds in its three-dimensional form, the homeowner will feel the thrill of creation – and with it an almost irresistible urge to fiddle with the design as the sunwing takes shape. For one thing, the addition will initially seem smaller than expected, an optical illusion peculiar to construction that makes a framed-in room seem half the size of the same room with finished walls. Resist the impulse to make major changes at this late stage: it will only wreak havoc with the budget and sorely test relations with the builder. Have enough confidence in the thoroughness of the plan to enjoy watching the sunwing emerge.

Construction will bring with it several weeks or months of disruption and disorder. The homeowner can keep this to a minimum by providing handy disposal bins for waste and by planning ahead where materials should be unloaded so that they are convenient yet not underfoot. To keep the mess confined, open the common wall at the last possible moment, and close it back up with doors and windows as quickly as humanly feasible. Remove drapes, carpets and valued furnishings from the adjoining room until the dust settles.

One final word: do not move in until the addition is finished. Once the furniture and plants are installed, there is less incentive to varnish the trim and seal the tile. Beg or borrow the money if necessary, take an extra week off work before the snow flies, but finish the sunwing before it becomes a permanent part of the house.

This advice is offered, knowing that in most cases, it will be ignored. The warmth and glow of the sun is seductive, and from the moment the shell is closed in, the family will be drawn to the sunny space: Mary Coyle could hardly wait until the glass was installed to move in her plants, and the Powells were enjoying "hot tub breaks" between sprees of construction long before the sunwing was complete. And when winter set in, they discovered, like other happy sunwing owners, that they had not only built an addition, they had created an oasis of invincible summer.

141

Sources

'Knowledge is of two kinds. We know a subject ourselves, or we know where we can find information upon it.'

— Samuel Johnson

DESIGNERS/BUILDERS

There are many firms and individuals across the country who offer passive solar design and energy-efficient construction assistance. Consult the yellow pages or the local chapter of the Solar Energy Society of Canada Inc. (SESCI).

Ron Alward
Memphremagog Community
Technology Group
RR 4
Mansonville, Quebec J0E 1X0
(514) 292-5563

David Bergmark
Solsearch Inc.
Box 1869, 34 Queen Street
Charlottetown, Prince Edward Island
C1A 7N5
(902) 892-9898

Michael Dorgan, Consultant
41 Rockcliffe Way
Ottawa, Ontario K1M 1B4
(613) 745-2948

John Hix, Architect
282 Borden Street
Toronto, Ontario M5S 2N6
(416) 533-5058

Richard Kadulski
The Drawing-Room Graphic Services Ltd.
1269 Howe Street
Vancouver, British Columbia V6Z 1R3
(604) 689-1841

Michael Kerfoot
Sunergy Systems Ltd.
RR 2
Carstairs, Alberta T0M 0N0
(403) 637-3973

Brian Marshall, Robert Argue
Sun Shelters
334 King Street East, Suite 208
Toronto, Ontario M5A 1K8
(416) 364-1044

Gerry Recksiedler
Down to Earth Solar Works
Box 744
Stonewall, Manitoba R0C 2Z0
(204) 467-8518

Don Roscoe
RR 1, McGrath's Cove
West Dover, Nova Scotia B0J 3L0
(902) 852-3789
Also sells Ventomatic solar vent opener imported from Britain and max-min thermometer; mail order.

R. Edgar Scrutton
Sun Parlor Solar Ltd.
Box 74A, RR 4
Amherstburg, Ontario N9V 2Y9
(519) 736-5815

Alan Seymour, Architect
136 Arlington Avenue
Toronto, Ontario M6C 2Z1
(416) 653-2193

Wayne Wilkinson
7 Ketza Road
Whitehorse, Yukon Y1A 3V3
(403) 667-4933

BOOKS

Butti, Ken and John Perlin, *The Golden Thread.*
Cheshire Books
Palo Alto, California, 1980.
A history of the development of solar energy from prehistory to the present.

Canadian Wood Frame House Construction.
Canada Mortgage & Housing Corporation
Ottawa, 1981
An indispensable reference on standard Canadian construction techniques.

Klein, Miriam, *Horticultural Management of Passive Solar Greenhouses in the Northeast.*
Newport, 1980
The Memphremagog Community
Technology Group
Box 456, Newport, Vermont 05855

Craft, M.A., (ed.), *Winter Greens: Solar Greenhouses for Cold Climates.*
Renewable Energy in Canada
Toronto, 1983
A step-by-step guide to the design and construction of an attached solar greenhouse, with a good section on management techniques.

Hix, John, *The Glass House.*
London, England 1974
A complete history of artificial growing environments.

Energy Efficient Housing Construction.
Canada Mortgage & Housing Corporation
Ottawa, 1982
This free booklet describes construction details for an energy-efficient house.

Marshall, Brian and Robert Argue, *The Super Insulated Retrofit Book.*
Renewable Energy in Canada
Toronto, 1981
Excellent advice on making homes energy efficient.

Nearing, Helen and Scott, *Building and Using our Sun-Heated Greenhouse.*
Garden Way Publishing
Charlotte, Vermont, 1977
The pioneers of the back-to-the-land movement describe their simple greenhouse attached to a garden wall.

Nisson, J.D. Ned and Gautam Dutt, *The Superinsulated Home Book.*
John Wiley & Sons
Toronto, 1985
Excellent up-to-date construction details for building an energy-efficient home, many based on the studies of the Saskatchewan National Research Council. Many techniques can be adapted to sunwings.

Schwolsky, Rick and James Williams, *The Builder's Guide to Solar Construction.*
McGraw-Hill
Toronto, 1982
An easy-to-read guide to solar-oriented construction. Includes some information specifically on sunwings.

Garden Way Manufacturing Company

Seymour, Alan and Michael Dorgan, *Construction Details for Attached Sunspaces.*
National Research Council
Ottawa, to be published
Replaces *The Solarium Workbook;* contains a comprehensive source listing of sunspace products and manufacturers.

Wolfe, Delores, *Growing Food in Solar Greenhouses.*
Dolphin Books, Doubleday & Co.
Garden City, New York, 1981

Check local agricultural colleges and the federal and provincial departments of agriculture for information on greenhouse horticulture.

PERIODICALS

Fine Homebuilding
The Taunton Press
52 Church Hill Road, Box 355
Newtown, Connecticut 06470

Harrowsmith
Camden House Publishing Ltd.
Camden East, Ontario K0K 1J0

New Shelter
Rodale Press
Emmaus, Pennsylvania 18049

Organic Gardening
Rodale Press
Emmaus, Pennsylvania 18049

Sol
Solar Energy Society of Canada
209-135 York Street
Ottawa, Ontario K1N 5T4

Solar Age
Solar Vision Inc.
Church Hill
Harrisville, New Hampshire 03450

INFORMATION/COURSES

The organizations listed below offer information on energy-related topics, either free or at nominal cost. Some also sponsor workshops and courses in energy-efficient construction and sunwing design. Write for further information.

Brace Research Institute
Macdonald Campus, McGill University
Box 900
Ste-Anne-de-Bellevue, Quebec H9X 1C0

Canada Mortgage & Housing Corporation
(CMHC)
Montreal Road
Ottawa, Ontario K1A 0P7
Or contact the nearest regional CMHC office
listed in the blue pages of the telephone book
under Government of Canada.

Canadian Home Builders' Association
(CHBA)
20 Toronto Street, Suite 400
Toronto, Ontario M5C 2B8
CHBA runs workshops on energy-efficient
construction that are open to homeowners.
Consult regional offices for details on
publications, courses and a list of trained
builders in your area.

The Institute of Man and Resources
49 Pownal Street
Charlottetown, Prince Edward Island
C1A 3W2

The National Centre for Appropriate Technology
Box 3838
Butte, Montana 59071

Solar Energy Society of Canada Inc.
(SESCI)
135 York Street, Suite 206
Ottawa, Ontario K1N 5T4
Publishes the *Canadian Solar Directory*, a
listing, by province, of businesses,
professionals and information agencies
involved in solar energy and other renewable
energy resources.

Solar Greenhouse Course
Department of Mines & Energy, Pilot Projects
Box 668
Halifax, Nova Scotia B3J 2T3
This seven-lesson slide-tape show with
manual, developed by Mines & Energy and
Solar Nova Scotia, is available for sale.
Trained instructors offer the course
throughout the Maritimes.

Ecology House
12 Madison Avenue
Toronto, Ontario M5R 2S1
Its library is open to the public.

Edmonton Energy Conservation Centre
10511 Saskatchewan Drive
Edmonton, Alberta T6E 4S1
Publishes *The Environmental Network
Newsletter*, energy news and views from
Canada's west.

Canadian Wood Council
85 Albert Street
Ottawa, Ontario K1P 6A4
Publishes *The Construction Guide for Preserved
Wood Foundations*.

L'Institut GRACE
3937 rue Berri
Montreal, Quebec H2L 4H2

Total Environmental Action Foundation Inc.
Church Hill
Harrisville, New Hampshire 03450

Check provincial departments of housing
or mines and energy (conservation and
renewable energy branch) for information on
energy-efficient construction and attached
greenhouse/sunspace design. As part of the
CREDA programme (Conservation and
Renewable Energy Demonstration
Agreement), many provinces funded the
construction of sunwings that can be visited
by the public.

PLANS/DESIGN AIDS

Attsun Sun Space Systems
The Old School
Hyde Park, Ontario N0M 1Z0
Blueprint ($10) for 9-by-12-foot sunwing with
angled south glazing (plastic), and solid
insulated roof and end walls. Can be adapted
to larger sizes.

Total Environmental Action Foundation, Inc.
Church Hill
Harrisville, New Hampshire 03450
Plans ($9.95) for a 9½-by-15-foot solar
greenhouse with 50-degree sloped south
glazing (glass) and insulated roof and end
walls.

Drawing-Room Graphic Services Ltd.
1269 Howe Street
Vancouver, British Columbia V6Z 1R3
Offers several designs for both attached and
freestanding solar greenhouses ($25).

Ecology House
12 Madison Avenue
Toronto, Ontario M5R 2S1
Plans ($10) for a sunwing with sloped south
glass and solid roof with skylights. Can be
adjusted for any size sunwing.

Solar Applications and Research Ltd.
3683 West 4th Avenue
Vancouver, British Columbia V6R 1P2
Plans ($25) for an attached greenhouse with
exterior hinged reflective flap over roof
glazing.

Sunflake Sun Room
Richard Feeney
Box 1698
Durango, Colorado 81301
Plans ($15) for a sunwing with vertical south
glazing and a solid insulated roof with
skylights.

Robert Tinker

Design Works, Inc.
11 Hitching Post Road
Amherst, Massachusetts 01002
Sells House Building Kit to help amateurs
build a scale model of interior and exterior of
an addition up to 1,500 square feet; Drawing
Kit that helps laymen produce accurate
detailed drawings of a building project; and
Solar Card that shows sunpaths.

PRODUCTS AND MANUFACTURERS

GLAZING

Check the yellow pages under Glass
Manufacturers and Plastic Products for local
sources of glazing, or write to the head offices
below for the nearest local supplier.

Chemacryl Plastics Ltd.
360 Carlingview Drive
Rexdale, Ontario M9W 5X9
(800) 268-4709
Acrylite SDP and Cryolon SDP, double-
skinned acrylic and polycarbonate rigid
glazings.

Rohm and Haas Canada Inc.
2 Manse Road
West Hill, Ontario M1E 3T9
Plexiglass acrylic plastic sheet and Tuffak-
Twinwal double-wall polycarbonate.

General Glass International Corporation
542 Main Street
New Rochelle, New York 10801
Manufactures Solakleer low-iron glass.

Graham Products Ltd.
Box 2000
Inglewood, Ontario L0N 1K0
Excelite, FRP, Exelac UV-coating and
accessories for fibreglass.

Polygal Plastic Industries
Box 272
Edgerton, Wisconsin 53534
Double- and triple-wall polycarbonate
glazings; plastic and aluminum mounting
extrusions.

Energy Alternatives Ltd.
95 Victoria Street
Amherst, Nova Scotia B4H 4B8
Glazing materials and accessories.

Hordis Brothers, Inc.
523 Forest Rose Avenue
Lancaster, Ohio 43130
Manufactures Heliolite low-iron glass.

Solar Components Corporation
Box 237
Manchester, New Hampshire 03105
Sun-Lite FRP systems and components;
water-storage tubes.

The following firms manufacture sealed
insulating glass units using Heat Mirror low-
emissivity film.

Aluminum Building Products
9155 Langelier Boulevard
Montreal, Quebec H1P 3A3

Clear-View Insulating Glass
45 Fenmar Drive
Weston, Ontario M9L 1M2

Debonair Industries Ltd.
2758 Norland Avenue
Burnaby, British Columbia V5B 3A6

SKYLIGHTS

Architectural Plastics Limited
5165 Timberlea Boulevard
Mississauga, Ontario L4W 2S3

Hickey Plastics Ltd.
2000 Ellesmere Road
Scarborough, Ontario M1H 2W4

Repla Ltd.
482 South Road East
Oakville, Ontario L6J 2X6

Velux-Canada Inc.
16805 Hymus Boulevard
Kirkland, Quebec H9H 3L4

INSULATING WINDOW PRODUCTS

Canadian G.E.M. Industries
3060 Lakeshore Road West
Box 403
Oakville, Ontario L6L 1J2
Window Quilt; has distributors in most
provinces.

EnerShade
40 Norwich Street East
Guelph, Ontario N1H 2G6
(519) 821-8998
Mail order; sells materials and plans for do-it-
yourself thermal shades and pop-in shutters;
also does custom manufacturing.

Sunergy Systems Ltd.
RR 2
Carstairs, Alberta T0M 0N0
(403) 335-4988
SunSeal Insulating Curtain and other
conservation products.

PREFABS

Write to the addresses below for local distributors of these prefabs. Check the yellow pages under Solariums and Greenhouses for other manufacturers. Many of these companies also sell glazing systems and accessories, such as shade cloth, movable insulation panels, fans, heaters, louvred vents, max-min thermometers and thermostatic controls.

Advanced Energy Technologies Inc.
Box 387
Clifton Park, New York 12065
Zeroenergy Room, a low-energy modular system with low-iron, almost vertical glass and steel-cased urethane foam panels.

Brady and Sun
97 Webster Street
Worcester, Massachusetts 01603
Laminated-pine arch and double-glass LivingRoom.

Canadian Greenhouses
Box 160
Grimsby, Ontario L3M 4N6
Plastic prefabs and greenhouse accessories.

Diamond Structures Ltd.
28 Bentley Avenue
Nepean, Ontario K2E 6T8
Cedar and glass prefab with automated between-glazing insulation.

English Aluminum Greenhouses
506 McNicoll Avenue
Toronto, Ontario M2H 2E1
Distributes Four Seasons aluminum and glass or plastic greenhouses with curved eaves.

Equipment Consultants & Sales
2241 Dunwin Drive
Mississauga, Ontario L5L 1A3
Greenhouse equipment and supplies.

Evergreen Room
Hwy. 26
Burnett, Wisconsin 53922
Laminated-cedar arch and glass solariums.

Four Seasons Renovations & Solariums
2010 7th Avenue
Regina, Saskatchewan S4R 0J7
Manufactures Dream Room laminated-wood arch and glass solariums and distributes aluminum/glass greenhouses.

Garden Way Sun Room
Box 697 Victoria Station
Westmount, Quebec H3Z 3Y7
Laminated-pine and glass solariums.

Advanced Greenhouses Ltd.
440 Phillip Street
Waterloo, Ontario N2L 5R9
Aluminum and double-wall polycarbonate freestanding and attached greenhouses.

Attsun Sun Space Systems
The Old School
Hyde Park, Ontario N0M 1Z0
Distributes The Vegetable Factory, double-wall fibreglass or clear acrylic double-glazed greenhouses.

B.C. Greenhouse Builders Ltd.
7323 6th Street
Burnaby, British Columbia V3N 3L2
Aluminum and glass greenhouses.

Solarium Ltd.
1195 Principale
Granby, Quebec J2G 8C8
Aluminium and glass with curved eaves.

Solcan Ltd.
RR 3
London, Ontario N6A 4B7
Aluminum-clad cedar and double-wall polycarbonate greenhouses.

Harnois Industries Inc.
1044 Principale
St-Thomas-de-Joliette, Quebec J0K 3L0
Greenhouse systems and accessories.

Jacobs Greenhouse Manufacturing Ltd.
371 Talbot Road
Delhi, Ontario N4B 2A1
Aluminum and glass greenhouses.

Lord & Burnham Co. Ltd.
Box 428
325 Welland Avenue
St. Catharines, Ontario L2R 6V9
Metal and glass or plastic greenhouses with curved eaves.

National Greenhouse Co.
400 E. Main Street
Pana, Illinois 62557
Metal and wood-frame greenhouses; supplies and accessories.

Northern Greenhouse Sales
Box 1450
Altona, Manitoba R0G 0B0
Steel-frame and plastic systems; supplies and accessories.

Pella/Rollscreen Co.
Pella, Iowa 50219
Pella Sunroom with double glass on aluminum-clad wood frame.

McLarty Solar Fabricating Ltd.
Box 24
Ailsa Craig, Ontario N0M 1A0
Laminated-wood arch and glass solarium/greenhouse.

Moderco Solar Windows & Greenhouses Inc.
3400 Losch Boulevard, #34
St-Hubert, Quebec J3Y 5T6
Aluminum and glass systems with curved eaves.

Sunwrights Multinational Inc.
205 Pretoria Avenue
Ottawa, Ontario K1S 1X1
Clear Arch Greenhouse of laminated-cedar arch and double-wall polycarbonate.

Zytco Solariums Ltd.
6969 Trans-Canada, Unit 120
St. Laurent, Quebec H4T 1V8
Aluminum and glass greenhouses.

Note: A listing in Sources does not constitute an endorsement of the product or service listed.

Index

Plans

Plans

¼'' = 1 foot

Plans

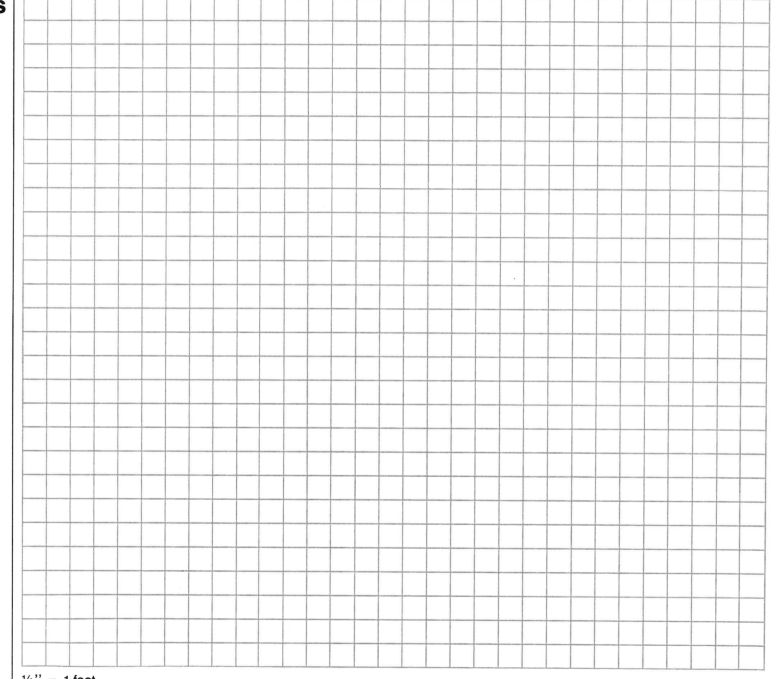

¼'' = 1 foot